Spy Plane

Spy Plane

INSIDE BALTIMORE'S SURVEILLANCE
EXPERIMENT

Benjamin H. Snyder

UNIVERSITY OF CALIFORNIA PRESS

University of California Press
Oakland, California

© 2024 by Benjamin H. Snyder

Library of Congress Cataloging-in-Publication Data

Names: Snyder, Benjamin H., author.
Title: Spy plane : inside Baltimore's surveillance experiment /
   Benjamin H. Snyder.
Description: Oakland, California : University of California Press,
   [2024] | Includes bibliographical references and index.
Identifiers: LCCN 2024011343 (print) | LCCN 2024011344 (ebook) |
   ISBN 9780520396029 (cloth) | ISBN 9780520396036 (paperback) |
   ISBN 9780520396043 (ebook)
Subjects: LCSH: Electronic surveillance—Social aspects—Maryland—
   Baltimore. | Police patrol—Surveillance operations. | Law
   enforcement—Technological innovations—Maryland—Baltimore.
Classification: LCC HV7936.T4 S62 2024 (print) | LCC HV7936.T4
   (ebook) | DDC 363.2/32—dc23/eng/20240618
LC record available at https://lccn.loc.gov/2024011343
LC ebook record available at https://lccn.loc.gov/2024011344

33 32 31 30 29 28 27 26 25 24
10 9 8 7 6 5 4 3 2 1

Contents

List of Illustrations *vi*
Preface *vii*
Acknowledgments *xi*

1 The Problem of Experimentation *1*

2 The All-Seeing Eye *27*

3 False Positives *57*

4 Experimenting on the Black Butterfly *87*

5 Big Brother's Bigger Brother *122*

6 Privacy and the Time Machine Problem *150*

7 Mechanical Witness *177*

8 No to Hype, Yes to Community Control *203*

Appendix: Watching the Watchers *227*
Notes *241*
Selected Bibliography *271*
Index *285*

Illustrations

FIGURES

1. How the spy plane works *3*
2. Analyst's track markings on spy plane footage *5*
3. Zoomed in on a shooter and victim *6*
4. Two views from the same spy plane footage: zoomed out on West and Central Baltimore; zoomed in on a street corner *29*
5. How a spy plane investigation unfolds *31*
6. Using a tracking report to pull up corresponding CCTV footage *39*
7. Location of original Video Patrol program, 2001 *95*
8. A Video Patrol monitoring kiosk, 1996 *97*
9. Inside the new Watch Center, 2005 *102*
10. Police announce CitiWatch expansion corridors, 2004 *104*

MAPS

1. Racial map of spy plane investigations, 2020 *66*
2. Tracking in the McHenry Street investigation *70*
3. Racial map of homicides by shooting in Baltimore City, 2020 *90*
4. Racial map of CitiWatch cameras *106*
5. Racial map of stingray use by Baltimore police, 2007–2014 *118*

Preface

In 2017, I began studying Persistent Surveillance Systems (PSS), a small tech start-up that would eventually conduct one of the most controversial mass surveillance experiments in US history. As a contractor for the Baltimore Police Department (BPD), PSS deployed an aerial surveillance system that, they claimed, could put the entire city under watch. Having endured almost a decade of over three hundred murders a year, and double as many shootings, Baltimore was desperate to turn things around and thought PSS's "spy plane," as the program came to be called by locals, could shift the tide.

In the winter of 2019, BPD entered into a contract with PSS. They asked the company to fly multiple planes above the city for a six-month trial, which was called the Aerial Investigation Research (AIR) pilot program. PSS's founder, Ross McNutt, invited me behind the scenes to observe. Then . . . disaster. The COVID-19 pandemic hit. We were all told to stay in our homes—"flatten the curve," "maintain social distancing." Yet, the spy plane program was still a go. Other researchers had been invited to check it out, but, I found out from McNutt, they were all staying home.

I'll never forget the moment. I'm sitting on my couch in Massachusetts next to my partner. She's three months pregnant with our daughter. I have been telling her all week that I should call off the project because it's too dangerous. What if I get sick? What if I bring the virus back home and get her, our son, and our unborn baby sick? Day cares had been shut down, too, and my partner would be on her own without any support while I was gone. We could bring up my mother-in-law, who, like the rest of my partner's family, is a Baltimorean, to help out, but this might put her in greater danger. The virus seemed to be even harder on older people.

After days of agonizing, we decided I should go. Looming large was the fact that there would be no one else to tell the story of what happened with the spy plane. Baltimore would never get a full accounting of how the city was used, once again, as a kind of petri dish to test out an experimental crime technology. My mother-in-law kindly offered her house to use as a base of operations so that I could more effectively quarantine while conducting the research. And she was willing to drive up from Baltimore so that she, my partner, and our son could be together. Every book is probably like this in some way, but the work in this study was even more so. Studying the spy plane up close required huge sacrifices and support from family and friends.

So, I went. I threw my packed bags, which had been sitting by the door all day, in the car. Even by the afternoon, I was still not convinced I would go. I hugged my son, not old enough to really know what was going on. We knew so little about the virus then, I barely knew how to better protect myself. There were no masks; at that point, we were told we didn't need them. I drove off, but I didn't get very far before I had to pull over. I stopped the car and began to sob while I took in what was happening. Was I really going to embed myself among police during a global pandemic?

The dilemma I felt in this moment was shaped by one of the core problems with contemporary high-tech policing. The panoply of sophisticated military gadgets that have been given to American police for the past several decades, such as facial recognition software, algorithmic decision-making tools, and gunshot detection systems, seem to continually disappoint. They haven't fixed the problems of the criminal legal system, unlocked new solutions to gun violence, improved trust in police, or even led to substantial increases in clearance rates for investigators. Yet police, city officials, and community leaders keep coming back. They are continually wooed by the magic of technology as a "solution."

One of the big barriers to understanding this dynamic is lack of access behind the scenes of today's high-tech policing market. Typically run through public-private partnerships, in which law enforcement agencies contract with for-profit firms that provide sometimes wildly untested prototypes, contemporary high-tech policing is incredibly opaque. We simply don't know a lot about how the companies that give these tools to police actually deploy them in a real-world situation. As I would come to find out, this lack of transparency is part of a much larger problem with the way police "experiment" on the public with untested tools.

So, here I was with unprecedented access to one of the most controversial experiments in mass surveillance ever, and also faced with the prospect of exposing myself, those in my field site, and potentially even my family to a dangerous new disease. But how could I back away from one of the few times a tech company had actually opened its doors so widely? What is more, the spy plane had flown once before in Baltimore, for a few months in 2016, in total secrecy from the public. No one had been there to fully document what happened back then. If I didn't go this time, that would happen again.

The fieldwork was fascinating but awful. Every night when I came home from the spy plane operations center, before I would spend hours writing up my field notes into the wee hours, I would clean and disinfect everything—wiping down car door handles, doorknobs, my phone, even my field notebook and pen. If there was virus on the objects I used to collect data, I didn't want it coming into the house and infecting me, or worse, infecting my mother-in-law when she returned to her home after this was all over. I washed my hands *constantly*, taking my temperature every time I felt a slight discomfort in my throat. But because of the stress and late nights, I often couldn't tell if my throat was sore because I was getting sick or because I was just so run-down. Before returning to my family after a bout of fieldwork, I would quarantine for the recommended period in my mother-in-law's house to try to make sure I didn't bring the virus back—completely alone for days with nothing to think about but all the violence I had witnessed through the strange, blurry footage of the spy plane.

It is not clear that I made the right choices here. I couldn't have known if I would get sick or pass the virus along (I didn't, as far as I know), or just how much of a toll the stress would exert on my family (it was huge). What is clear is that if I had not gone, Baltimore would never know some of the things that went on that summer as the spy planes flew overhead. Much of what I discuss in this book has never been reported to the public. In the end, the experience completely changed the way I think about the promises and perils of high-tech policing.

Acknowledgments

It is a little strange to thank a surveillance company in a book critical of surveillance. The truth is I owe a huge debt of gratitude to Persistent Surveillance Systems, particularly Ross McNutt. It is rare that a tech company is as transparent as McNutt's. The way he and his team allowed me to directly observe and provided all internal documents necessary for the research should serve as a model for other companies.

I received so much support from folks in Baltimore. Joyous Jones was a constant source of knowledge in trying to give voice to the community supporters of the program. She treated me like family from the moment I met her and taught me so much about Baltimore. Dan Staples, of Open Justice Baltimore, helped me navigate the Maryland court records system, so I could keep track of prosecuted spy plane cases.

Brandon Soderberg walked with me through every stage of the research and writing. He even fact-checked the final manuscript, without me having to ask. Every ethnographer should be so lucky to have a crack investigative journalist in their corner as a collaborator.

Students at Williams College, too numerous to count, read early drafts and helped me navigate many of the ethical dilemmas of the fieldwork. Sneha Revanur helped shape the final manuscript for an undergraduate audience.

I am also indebted to a long list of colleagues who provided feedback. The Institute for Advanced Studies in Culture at the University of Virginia provided space and funding to begin the fieldwork. The Data and Society Institute served as a first audience to test out these ideas. Conversations at the Oakley Center at Williams College helped me discover what this book was about. Cory Campbell and Drake Reed helped me learn GIS and craft the maps in the book. Bret Anderson, Sarah Brayne, Sorcha Brophy, Joe Davis, Alison Gerber, Zeynep Gürsel, Jason Hill, Joe Klett, Ekédi Mpondo-Dika, Jim Nolan, Allison Pugh, Harel Shapira, Olga Shevchenko, Ben Shestakofsky, Christina Simko, Forrest Stuart, Julia Ticona, Francesca Tripodi, and Robin Wagner-Pacifici all provided crucial advice at different points in the research.

Greta Snyder read and helped me think through every word of every chapter. Her inexhaustible support and belief in this project (during a pandemic, no less!) literally made this book possible. Elliot, Liv, and I sure are lucky.

1 *The Problem of Experimentation*

Sarah sits me down at her analyst terminal inside the spy plane operations center. She pulls up a file of some cell phone video confiscated by detectives at the scene of a homicide earlier that week. The incident occurred on a street I will call "Patterson Street" in a West Baltimore neighborhood that sees dozens of shootings and murders a year.[1] It was the middle of the day, right in the middle of a neighborhood barbecue. Before clicking play, she warns me the footage is pretty grim. "There is some screaming and it's kind of disturbing," she says, and hands me a pair of headphones so she won't have to listen to it again. From a tight-knit orthodox Jewish community in Northwest Baltimore, she has not been exposed to violence like this before.

The video starts and I see the inside of a house. The camera is flailing back and forth, so I don't see much at first, but I hear a commotion. There is screaming from someone off camera.

Wait. I need to pause here.

I could go on and describe to you all the gruesome details I saw in this video, but something's off about that. I don't have permission from the person who made this video—a young Black woman from one of Baltimore's deeply segregated and hyper-policed communities, who is a relative of the man who was shot—to share these details.

Detectives took the footage from her and gave it to Sarah's team. I'm not even sure the video was given up freely. As I'm watching the footage of this woman's loved one being shot and bleeding out on the pavement, I'm thinking about this: "How are civilian employees of a private, for-profit surveillance company allowed to see this?" It feels so personal. I'm not going to recount this horror here. Instead, I will give you this: the text equivalent of image blurring.

The video ends—all sixteen seconds of it.

Sarah and I sit in silence for what seems like minutes, but is probably seconds, not knowing what to say. Finally, she grabs her mouse and slides the video back a few seconds. "At first I couldn't concentrate on anything because of the screaming and the body," she says, "but once that wore off I was able to start going through it frame by frame. You would be amazed how much you can pull out." Moving back to the few seconds after the sound of a gunshot, as the camera pans wildly, she points out two cars parked in the background at odd angles, blocking an intersection. You can just barely see the shooter getting into one car and his accomplice into another before they drive off, the edges of their frozen, ghostly figures streaked by the speed of the camera's panning.

Sarah then pulls up imagery captured that same day by Baltimore's Aerial Investigation Research (AIR) program, a first-of-its kind mass surveillance experiment brought into the city in 2020 to curb violent crime. Persistent Surveillance Systems (PSS), the small tech start-up that operates this program as a private contractor for the Baltimore Police Department (BPD), refers to this technology as wide-area motion imagery (WAMI). In Baltimore, they call it the spy plane.

The spy plane system creates a second-by-second, moving image of an entire city. The imagery is of such high resolution that you can zoom in on individual cars and people and track their movements for hours at a time. Because the imagery is saved on big hard drives, you can also "fast-forward" and "rewind" the movements of the city,

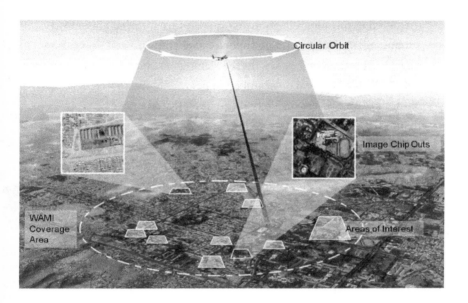

FIGURE 1. How the spy plane works. (*Source:* Persistent Surveillance Systems. Used with permission.)

watching how things played out over and again. The company's founder, Ross McNutt, describes it as "Google Earth with a TiVo capability." Police tasked McNutt's company to use this technology to help their detectives track suspects and witnesses in violent crime cases, in the hopes that it would increase case closure rates. They also hoped for it to deter crime—to send a message to Baltimore's "trigger pullers," as they're sometimes called, that police are watching, so don't do the crime.

Here's how it works. As illustrated in figure 1, an unassuming, small-engine, piloted aircraft is equipped with twelve 192-megapixel cameras. (It is not a drone, which is very expensive to run and too light and unstable to create usable imagery.) The cameras point out the side door of the plane, angled slightly downward. As the plane flies in a circle over the city, at roughly five thousand to ten thousand feet, it takes one gigantic photo every second, capturing about thirty square miles of terrain. Using sophisticated software and processing developed by PSS engineers, these images are stitched together into video-like footage and downlinked in near real time to a remote operations center where the images can be analyzed by a team of investigators. PSS archives all the footage on gigantic, secured hard drives inside the operations center, which can be accessed quickly and repeatedly by dozens of analyst terminals all at the same time. Each image is given a precise time stamp, down to the second. This allows multiple analysts to simultaneously look not just at the movements of the city right now, but also "go back in time" and look at movements recorded hours, days, weeks, or even years ago. Observers of the technology have compared it to a time machine.[2]

As the image refreshes, second by second, trained analysts like Sarah can pick out things moving on the ground and track them. As illustrated in figure 2, they use proprietary software, called iView, to place colored circles and lines on people and cars found in the imagery to convey their "pattern of life." This tracking is supposed to help detectives identify leads and, in the best-case scenario, point police to where a suspect can be intercepted and arrested. Put two or three of these systems in the sky, as they've done in Baltimore, overlapping their orbits at the edges to prevent gaps, and you can watch an entire city. Or, at least, that's how it's supposed to work.

As I'm sitting with Sarah, I begin to see how it really works. She shows me how her team searched through the spy plane footage to find the Patterson Street shooting suspects and track them backward in time. Analysts wanted to figure out what the suspects were doing before the incident to see if they could help police identify them. Using iView, Sarah "fast forwards" the spy plane footage until it matches the exact moment of the shooting recorded on the cell phone video's time stamp, then zooms down with her mouse wheel into the specific intersection of the incident captured on the cell phone video. Having

FIGURE 2. Analyst's track markings on spy plane footage from Ciudad Juárez.

synced the two sources in time and space, she tells me how her team of analysts matched the layout of the cars in the phone video with the layout of cars captured by the plane's cameras. She points out the two cars—little grayish pixelated smudges, about the size of a grain of rice on the screen—sitting at precisely the same odd angle as in the cell phone video. If you look hard, you can even see tiny little dots that might be the shooter and victim. "Here they are, see?" Sarah says, pointing. Using iView, colored circles were placed over the cars, and labels were added that said Vehicle 1 and Vehicle 2.

You have to see the footage to really understand how painstaking this tracking work can be. Spy plane imagery is hard to decipher by the untrained eye. Many people say, at first glance, it looks like an X-ray or ultrasound. Because the BPD would not allow me to take photos of spy plane imagery in Baltimore, however, the best I can do is show you imagery from some of the only footage that has been released to the public. Figure 3 shows a single frame from a murder in Ciudad Juárez, Mexico, where PSS operated briefly in 2009. The image is at

THE PROBLEM OF EXPERIMENTATION [5]

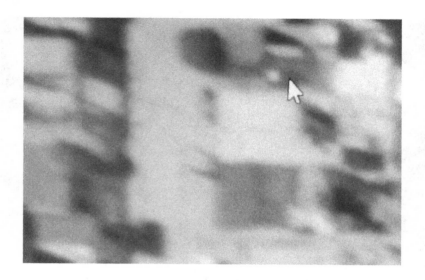

FIGURE 3. Zoomed in on a shooter and victim in Ciudad Juárez.

maximum zoom. The short, slightly downward sloping dark lines in the middle of the image are the shadows of the shooter and victim cast on the ground by the afternoon sun right before the trigger is pulled. Zoomed in as far as possible, then, an individual on foot appears on an analyst's monitor as roughly one pixel in size—like a smudgy dot against the ground. The way Sarah can quickly ping back and forth between high-resolution ground footage at the Patterson murder and the grainy overhead spy plane footage of the same scene, and understand what matches what, is a product of hours of training that attunes one's eyesight to the tiniest pixel-level details. That's why McNutt sometimes jokingly refers to tracking work as "blob-ology."

Sarah continues to lead me through her process. Because the spy plane captures imagery of almost the entire city, her team was then able to zoom out from the level of the crime scene and "rewind" the footage. This allowed them to watch the movements of the cars on Patterson Street, before the time of the murder, well beyond the limited view of the cell phone video. Frame by frame, click by click, covering several hours before the incident, they tracked the two cars past dozens of public CCTV and private security cameras. Detectives would have never thought to check these cameras on their own unless the spy plane had pointed them there. Analysts pulled higher resolution footage of the suspects' appearance and behavior, feeding it to Detective Thomas, the lead investigator on the case.

Ultimately, they identified seven people involved in the murder. Four were not seen in the original cell phone video because they were out of the frame of the lens, but analysts were able to connect them to the crime scene by tracking their cars. PSS analysts tracked all of these people to multiple locations all over West Baltimore. They even have imagery of the men repeatedly driving past the eventual crime scene three separate times, hours before they finally returned to do the job.

This *pattern-of-life* analysis, as it is called in military intelligence, was summarized in a final briefing document and handed over to police. PSS concluded that the imagery "documents a 40-minute coordinated and premeditated plan." A massive packet of evidence was put together to supplement Detective Thomas's other work. Meeting Thomas a few days later, I ask him if the aerial footage helped the case. "Yeah, I basically would have had nothing if it wasn't for this," he said. "This gave me my whole case."

The Patterson Street case represented precisely the kind of violence PSS had been brought into the city to address. Where violent crime had been declining all over the United States in the past decade, Baltimore had been an outlier. After a small decline through the early 2000s, Baltimore's murders and shootings had increased to record levels by 2020. City leaders were desperate for a solution and they decided to try out an experimental program that had never been deployed on so large a scale in a US city. As a hungry start-up, PSS dreamed of being the "unicorn" tech firm that could make Baltimore safer using one of the US military's experimental tools.

In the end, the work on the Patterson Street case amounted to nothing. By week's end, Thomas had to all but abandon the case because of an unexpected problem: the prosecutor's office refused to prosecute. Despite analysts developing a thorough mapping of the suspects' movements, which tied their actions in the cell phone video to addresses, vehicles, partial license plate numbers, and other partially identifying information, the prosecuting attorney assigned to the case was unconvinced. It was all circumstantial. Prosecutors told Thomas that unless he could provide evidence that directly and unequivocally identified the men in the video by name, they would not proceed with the case, should he make an arrest. Thomas was more or less forced to drop the case, his attention getting pulled to the four other homicide investigations he had been thrown just that week.

Far from being seen as a powerful Big Brother–like, all-seeing eye, to many inside the criminal legal system of Baltimore, the spy plane became more like a Big *Bother*. In June 2021, the technology was deemed unconstitutional by a federal court—an outcome that many privacy advocates warned would happen. A new mayor was elected who called the program a "gimmick."[3] City leaders abandoned the program and grounded the planes for good. Years later, all the court cases brought based on spy plane evidence were thrown out. The victims' families and the few arrested suspects (some of whom spent years in jail awaiting trial) had their whole lives upended by this program and were now left to fend for themselves and pick up the pieces. The spy plane experiment was a failure.

This book is the only behind-the-scenes account of how this controversial program unfolded. Between 2017 and 2023, I was allowed total access to the inner workings of the program and given an unprecedented look at how a small start-up develops and deploys a new tool in the exploding market for law enforcement technology. At a time when police are spending millions of dollars on fancy new gadgets that purport to turn officers into something like supercops, I came away with a far more complex picture.

The main thing I found is something I call the "problem of experimentation." The scientists, engineers, businesspeople, law enforcement, and elected officials who create and deploy crime technologies seem to be fine with treating some neighborhoods, particularly race-class subjugated neighborhoods, as spaces to conduct radical experiments with potentially dangerous prototypes.[4] When police experiment, they often don't even have to tell the public they're doing it or report to the public if the experiment has borne fruit. When the tools underperform, break, are abused, make mistakes, or result in unforeseen consequences that hurt people's lives, rarely is anyone held to account. In part, this is because the harm has often been shouldered primarily by society's most marginalized citizens. Much later, I came to realize that this kind of experimentation isn't just limited to the crime technology market, either. It is, in fact, a core part of the "move fast and break things" culture of many contemporary tech start-ups.[5] This book is about what is wrong with that culture of experimentation and how we could do things better.

The Boomer/Doomer Hype Cycle

To understand why I focus on experimentation, first you have to understand how the public debate around new technologies usually plays out. This was not the first time the spy planes had flown in Baltimore. PSS came to the city in 2016, under total secrecy. WAMI had also been deployed briefly in a handful of North American cities (Juárez, Dayton, Compton, and Philadelphia). All the accounts of these early experiments led me to expect something out of George Orwell's dystopian novel 1984. Time and again, opinions from the popular media, and even a lot of the academic scholarship, described the spy plane as the embodiment of either a dystopian nightmare (in which the state is invading our privacy) or a utopian dream (in which police could, as the company's tagline goes, "solve otherwise unsolvable crimes"). I came to find out that this polarized debate between techno-utopians and dystopians is not unique to the spy plane. It is common to our public language for metabolizing a variety of so-called disruptive technologies. I call it the "boomer/doomer hype cycle." This is what I take aim at in this book. I want us to look at things in a different way.

Having its roots in the early days of Silicon Valley, tech "boomerism" is associated with the '60s countercultural communalism of people like Steve Jobs, who were the first to cash in on the "tech boom" of the 1990s.[6] While they began with the baby boomer generation, tech boomers can be found among the young and old alike. In the realm of law enforcement technology, boomers often come from the tech industry, law enforcement, and pro-police citizens and government officials. For them, technology is merely a "tool in the toolbox" to aid police in doing their work faster and more effectively. In this way, boomers often assume that the state is infallible—it is a competent, sophisticated force for good in the world.[7] Technology will just enhance this power. Why wouldn't you want to help these benevolent actors save more lives, boomers ask, by giving them the best tools? If you're not doing anything wrong, they stress, you have nothing to fear.

Undergirding boomers' views is an ideology that scholars call *solutionism*—the notion that technology can "fix" persistent social problems without having to engage the deep, institutional, root causes of those problems.[8] Eric Adams, the former NYPD cop who took seat as the mayor of New York City in 2021, embodies boomer solutionism particularly well. In a press conference to tout his new "Blueprint to End Gun Violence," for example, Adams complained,

It blows my mind how much we have not embraced technology, and part of that is because many of our electeds are afraid. Anything technology they think, "Oh it's a boogeyman. It's Big Brother watching you." . . . No, Big Brother is protecting you.[9]

Adams's statement is surprising because the NYPD is one of the most equipped and expensive police forces on the planet. As political rhetoric, though, it is effective. Under the cover of a rise in crime statistics, themselves dubious, Adams invested heavily in notoriously glitchy technologies, like facial recognition cameras, as well as highly experimental things, like "AI" software that can supposedly detect a concealed gun.[10] After a high-profile shooting on a Brooklyn subway car in 2022, for example, Adams pacified public panic by pointing to this new untested tool. "We're going to explore new technology to make New Yorkers safe," Adams assured, "and we believe we have a technology that we can use in a subway system . . . that could detect weapons, and we're excited about the possibilities."[11]

Boomers are often obsessed with the feeling of security. In fact, how technology makes people feel is sometimes more important to them than whether or not it works. From their point of view, the problem is that "people" (it is never specified which people) don't "feel safe." If adding some new gadgets to the environment can change the emotional temperature of civic space, that's seen as an immediate solution and much easier than trying to address the root causes of people's insecurities.[12] As I discuss in this book, technological solutionism can be incredibly effective at soothing public worries, especially among people who live amid constant gun violence. US culture in the twenty-first century, following decades of hype from places like Silicon Valley, elevates technology to an almost sacred status. It is no wonder, then, that the citizens most at risk of dying from gun violence would feel attracted to the utopian dream that some powerful new tools could provide an immediate feeling of safety.

On the side of doomers, you typically have privacy activists and advocates, such as the American Civil Liberties Union and Electronic Frontier Foundation, as well as much of the popular media that focuses on technology, such as *Wired* and *Popular Science*, and the technology sections of the major news outlets. Doomers are often dystopian. They concoct sometimes preposterous scenarios about the "existential threat to all of humanity" of emerging technologies.[13] Police and other government actors, though seemingly tasked with providing safety, actually represent this kind of existential threat because of how the state's gaze invades everyone's privacy. Whatever an invasive technology does in terms of improving safety, it violates certain social norms about the exposure of personal information, which typically registers in our individual emotional reactions as a kind of nauseous anxiety. That's why it's bad. The state's penetrating gaze makes us feel violated, out of control, dirty—like an unwanted touch from a threatening stranger. Policing technology is not a benign tool, then, but a creepy weapon.

Often invoked is a comparison to the panopticon—the infamous (and creepy) circular prison envisioned by the nineteenth-century philosopher Jeremy Bentham and popularized by social theorist Michel Foucault.[14] *Pan* means all and *opticon* means something that sees. An article from the news site *Slate* gives an idea of how this metaphor is used to stoke dystopian fear. Titled "Very Little Stands between the U.S. and a Technological Panopticon," the author notes that "technologies such as license plate readers, 'geofencing,' . . . and especially facial recognition . . . [are] giving law enforcement the ability to build astonishingly detailed portraits of people's lives."[15] As far-reaching as these technologies are now, the article claims, their power is only growing: "Improvements have been exponential . . . and near-flawless facial recognition technology is likely within the industry's grasp." What you might notice is that behind the dystopian fear here is a form of hype—what historian of technology Lee Vinsel has cleverly called "criti-hype," or criticism that "both feeds and feeds on hype."[16] In critiquing technologies as "Orwellian," "Big Brother," or "panoptic," doomers are actually puffing up the public image of the companies that engineer them by making the technologies sound more powerful than they may be in reality.[17] Criticism turns into hype by restating unfounded information about a tool's power.

The boomer/doomer frame creates a cycle that overhypes new technologies, simultaneously stoking too much optimism and too much fear, such that we never actually get to have a clearheaded debate. Interest in the new gadget quickly wanes as we get used to it, or it dies out because it never reached the heights that were promised. Then the next gadget comes along, and we do it all over again.[18] This cycle is bad for the public, I argue.

The boomer/doomer hype cycle is fueled by a number of key assumptions that are actually shared by the two sides. Foremost, it assumes that relatively untested and seemingly magic technologies actually work as advertised. That's a really big assumption. Many of these tools have never been deployed in a real-world situation. In fact, when police deploy these technologies, *they are often actually helping the tech companies field test them for the first time*. Even the inventors might not really know if they work.

A second assumption is that most personnel within the criminal legal system (whether you see them as fallible or infallible) want to and are *able* to use new surveillance technologies. Driving the new regime, both sides seem to think, are police who are eagerly deploying these new gadgets, prosecutors and judges who will gladly welcome them into the courtroom, and third-party tech companies who understand and are able to integrate themselves into established patterns. Yet, criminal legal bureaucracies, like all bureaucracies, are incredibly complex ecosystems of conflicting and competing interests, full of flaws, biases, infighting, and entrenched habits. Why would we expect any new tool to just "work" when dropped into this kind of organizational chaos?

Lastly, both boomers and doomers (though, less so for the latter) tend to ignore or downplay inequality. When doomers talk about how new technologies are "watching us all," or when boomers proclaim that these tools will "make us feel safe," they rarely specify the "us." They gloss over the fact that the state's supposedly panoptic eyes are typically unequally focused on poor communities of color, queer folks, religious minorities, and other groups who have been pushed to the margins of society. In fact, especially when these technologies have never been deployed in a real-world situation, they are often systematically flawed or biased in ways that disproportionately affect marginalized groups.[19]

The Patterson Street case, and many other things I observed about the spy plane, made me question the costs of debating technology within the boomer/doomer paradigm. What if the technologies don't actually work all that well? Even if they "work," what if police departments are so dysfunctional that they can't make them work reliably and safely? If the criminal legal system is not, in fact, a unified system, but a conglomeration of relatively autonomous subunits with competing agendas, what if tech companies cannot figure out who the technology is for and how best to deploy it? Given the bureaucratic mess that characterizes most municipal policing and legal bureaucracies, isn't it a bit naive to think of the kind of surveillance being conducted by police as all-seeing? And if today's high-tech cops are not actually like Big Brother, does that mean there's nothing wrong here? Or does all this glitchy, experimental tech and its sometimes blundering, ham-handed users have some costs to society that we can only see with a reorientation of our attention?

The Problem of Experimentation

The reorientation we need now, I suggest, is to take technology down off the altar of hype/criti-hype and look at it in its social context. Technology is not magic, nor does it change society in a simple linear way. In place of Big Brother, the panopticon, solutions, and fixes, we should use words like *experiment*, *prototype*, and *pseudoscience*. We should talk much more about things like hype cycles, glitches, biases, mysterious or hidden risks, and the unintended consequences that happen during deployment. We should turn away from the utopian hope of solutionism and the dystopian fear of creepy supercops and focus on what I will call the problem of experimentation.

Experimentation refers to an ethical framework that I think we should use for looking at so-called "disruptive" technologies. It begins with the fact that only certain publics, not "all of us," are typically used as sites of tests and trials for new tools. Scientists, engineers, businesspeople, law enforcement, and elected officials continually target race-class subjugated neighborhoods, using them as a kind of petri dish in which to conduct experiments with technology.

By experiment here, I don't mean a classical double-blind science experiment with a preexperimental ethics review, careful data collection, and a clinical trial period that documents risks. I mean pseudoscience.[20] I mean experiment in the looser sense of "trying out," "working out the kinks of," or "kicking the tires on" a new technology. (These phrases have literally been uttered by officials when deploying a new anti-crime tool.)

These experiments are often radically opaque and full of unknown risks. When the technologies don't work as advertised, and they rarely do, they often make mistakes that harm people, such as false arrests, unwarranted police contact, or failed prosecutions. These harms are unevenly shouldered by those who live in the race-class subjugated neighborhoods that were made into test sites. It is sometimes not until police have already tried out an experimental tool in dozens, or even thousands, of investigations, and harm has already been done, that the public is even made aware that the technology exists. What is more, these experiments are incredibly expensive, often drawing on millions in tax dollars, grants from public institutions, or gifts from unelected wealthy donors—money that could be spent on other proven techniques. And on top of all this, there is rarely a reliable external review that can tell us whether the technology was actually effective at reducing crime (or whatever the goal) such that we could weigh the social costs of the harms and risks that were discovered during the experiment.

My worry is that the problem of experimentation is too often overlooked because the clamoring between boomers and doomers focuses our attention on some "yet to come" or "just around the corner" tech future.[21] This leaves little room to articulate dangers that are happening right now, as the technology is being tested on often nonconsenting subjects. To be sure, I am worried about the long-term horizon toward which all this technology is headed, too, but the relentless focus on this abstract future takes up so much public attention that it pushes out concerns about the concrete harms of pseudoscientific experiments on people today.[22]

I'm not the only one—and far from the first—to have tried to get the conversation to focus on real harms happening now, instead of a utopian/dystopian future. The strongest voices to articulate this position have been activists, scholars, and engineers documenting what sociologist Ruha Benjamin has called "discriminatory design." These are design decisions that "encode judgements into technical systems," which are typically invisible to the designers in the development phase but become all too clear during deployment.[23] Promoting their tool as objective and neutral, engineers often become aware of these biases and flaws only after the tool has already harmed people.[24]

A growing number of researchers in the social sciences[25] and a few vocal engineers inside the tech world,[26] as well as activist groups,[27] have taken a deeply empirical and inequality-focused approach to understanding policing technologies. They have examined, among many tools, PredPol's predictive policing program, Palantir's database surveillance system, ShotSpotter's audio surveillance technology, Axon's ubiquitous police body cam service, and the numerous facial recognition companies vying for dominance. Time and again, scholars and activists have shown that these so-called "game changing" new technologies are more hype than reality. They are often glitchy, biased, unreliable, underperforming, and sometimes even downright dangerous, especially for the communities being used as test subjects.

Though this pathbreaking work has made considerable headway, perhaps the biggest barrier to its further development has been a lack of detailed evidence from concrete case studies.[28] Much of the research conducted by activists and scholars concerned with present harms has been indirect and empirically thin. This is not a criticism of their work. Rather, it reflects the incredible lack of transparency from both police and tech companies, which has forced researchers to rely on Freedom of Information Act (FOIA) requests, leaked documents, interviews with ex-employees and police, and other retrospective or secondhand accounts. That's a problem, because this kind of indirect evidence provides only a narrow window on what is surely a complex and messy process of deployment. Very few have been let in the door to just sit and directly observe how a new policing technology is deployed, and then follow its impact as its data passes through the many phases of the criminal legal system.[29] We still don't know much about how tech start-ups in the criminal legal space operate behind closed doors as they help police "work out the kinks" on a new gadget. Additionally, the FOIA and leaked documentation is produced by those inside of these programs (police, engineers, company PR people, etc.), rather than being gathered independently by an unaffiliated observer. That means it's hard for us to know just how representative these documents are of what tech companies and police are really up to. What about all the stuff that doesn't make it into an official document? What about the stuff that companies want to hide from the public record? Only an independent observer on the inside would be able to see that. That's where this book comes in.

Watching the Watchers

Though there are a few firsthand accounts of spy plane investigations, they are limited to a few days of poking around.[30] This book is the first to give an extensive explanation of how investigations involving the spy plane play out, from 911 call, to arrest, to court case. How did I get this access? In 2017, I heard an episode about PSS on the public radio podcast *Radiolab*.[31] Fascinated by the program, I asked the man behind the whole story if I could study his company, and he said yes. It was as simple as that.

Ross McNutt, one of the inventors of the camera and data systems behind the spy plane, is the founder of PSS. I spent a lot of time with him during the six years it took to research this book and got to know him well. He is in his late fifties and stands a little over six feet tall. A surprisingly kind, stubbornly positive, and often goofy personality, he does not resemble the militaristic character one might assume of someone who heads a creepy-sounding surveillance company. Heavier and with less hair now than in his old military photos, he moves and talks with an intense, bubbly energy. Often dressed in cargo pants and golf shirts, and always with a navy blazer at hand when he needs to look more professional, he speaks in alternating bursts of analytical excitement and a kind of aw-shucks informality, interspersing highly technical analysis with little phrases like "Oh, by the way . . ." and "We're gettin' there!" and "Overworked and underpaid!" Just about everyone who has met him is taken aback by how . . . well . . . sweet he is.

Another surprising thing about McNutt is his openness to outside observers. Unlike most start-up founders, he obsesses over creating "total transparency," as he often phrased it to me, for his company and his technology. I first approached McNutt at a rather down moment for the company—shortly after the spy plane had been shut down for the first time in Baltimore in 2017. I asked him if I could include his company as part of a study on technology and policing. He told me that he welcomed queries like this. A few days later, I was on a video chat with him, and he spent two hours walking me through the raw footage from one of the company's most successful investigations in Juárez. A few months later, I was at the company's headquarters in Dayton, Ohio, for three days and given a crash course in how to do spy plane analysis. Between 2017 and 2019, I got to know many of the major backers of McNutt's proposed program and conducted numerous interviews in Baltimore and St. Louis with them.[32]

Finally in 2020, after a long struggle to convince Baltimore leadership to let the company back into the city, McNutt secured a six-month contract to fly multiple spy planes at the same time, something that had never been done. As part of this deal, he agreed to allow a team of external evaluators to comb through the program's operations. These included the RAND Corporation, who would assess the program's effectiveness; NYU's Policing Project, who would assess its ethics and privacy implications; and a team from the University of Baltimore, who would track public opinion. Each of these teams has produced reports of their findings.[33] Because of my relationship with the company prior to 2020, PSS invited me to this evaluation team as a qualitative researcher. Because of restrictions on travel related to the COVID-19 pandemic, however, which struck just as the program launched, none of the other researchers could visit the operations center in person. I was the only person not employed by the company or BPD who was actually there for a substantial period of time.

I was given total unfettered access to everything. I was retrained as an analyst and given a key card that accessed the company's data terminals to look at raw spy plane footage for myself. In sociology, this method is called *participant-observation*. While I could watch spy plane analysts at work and interview them about their experience (the observation part), I could also use my direct access to the footage to reconstruct the tracking myself and see if I agreed with what analysts saw and said (the participation part).[34] I did usually agree, but sometimes I didn't. Contradictions between what I observed or heard people say and what I saw in the raw footage could be incredibly revealing.[35]

I also interviewed and observed a handful of BPD detectives who worked dozens of cases using spy plane footage, which allowed me to see how they integrated this new technology with more familiar tools, such as CCTV, facial recognition software, or automated license plate readers.

After the program was shut down, in late 2020, I was given access to all the company's investigative files, which allowed me to follow how the evidence produced inside the operations center traveled through the court system. For another three years, I listened over Zoom to court hearings, pulled court filings, interviewed prosecutors and defense attorneys, and sat through one jury trial to further understand how the spy plane interacted with the law, a process that extended well into 2023. In total, research for this book took around six years to complete.

Why did McNutt agree to let me in? In part, his motivation seems to come from his beliefs as a technological solutionist. "If people just see what it is we do," he would often say to me, "they won't be afraid." A true believer in technology's ability to fix social problems, McNutt promoted the spy plane as a way to "solve otherwise unsolvable crimes." If the public could just understand how the technology actually works, his thinking ran, they would cast off any worries about privacy or creepiness. McNutt seems to have seen me as a vehicle to broadcast that solutionist message by simply providing an accurate, independent, on-the-ground account. In this spirit, he stressed that he wanted my research to be independent. He never told me what to write or even asked me to write a favorable analysis of the company. I even signed a contract with the company that gave me "total academic freedom." I never took money from the company or its financial backers while doing the research. I describe the process of the fieldwork in greater detail in the appendix, but it is worth stating here that this was incredibly ethically fraught work. Though I am sure I made mistakes, and my point of view is as colored by my politics and social position as any other piece of social research, this book is the most level-headed and honest account of what I saw that I could muster.

Origins

The spy plane comes from the battlefields of Iraq. Holding a doctorate from MIT, Ross McNutt had a long career in the air force specializing in rapid product development (RPD). Where the air force had long followed the lumberingly slow model of institutions like the Defense Advanced Research Projects Agency (DARPA), with rounds of designing and testing before deploying nearly perfect technologies in the field, McNutt helped popularize the idea of "rapid prototyping."[36] Rather than wait for a technology to approach perfection in a controlled setting before unleashing it on the world, RPD resembles the "move fast and break things," "fake it 'til you make it," or "fast failure" mentality of Silicon Valley.[37] Engineers develop prototypes with both known and unknown flaws, immediately deploy them in the field on a small trial basis, and then rapidly redesign them, perfecting them as they go.

In the military, spy plane technology is called wide-area motion imagery (WAMI).[38] McNutt did not invent it single-handedly but was a big part of a group of teams that produced competing prototypes. Documented extensively by technology ethicist Arthur Holland Michel in his book *Eyes in the Sky*, the idea for WAMI arose in the context of the US military's drone program.[39] In an attempt to "police the world" during the war on terror by maintaining a "constant stare," the military developed aerial camera systems that can see persistently.[40]

Unmanned, remotely controlled drone aircraft became a huge part of this effort, but they quickly revealed a severe limitation: the "soda straw" problem. When drone operators look down with their precision cameras, their view of the ground is like looking through a drinking straw. This makes drones excellent at maintaining a constant view of a target that has already been identified, but limits their ability to find the target in the first place. If operators do not already know where to point the camera, finding a target with a drone becomes a needle-in-a-haystack operation. WAMI systems are meant to solve this problem. With access to a wide-angle video of everyone's movements on the ground, analysts can reverse the footage and zoom in on specific targets, once they realize they are of interest.

As has long been the case for military technology, Hollywood dreamed up WAMI before military engineers built it.[41] In the 1998 dystopian film *Enemy of the State*, a lawyer played by Will Smith is tracked across Washington, DC, by an NSA satellite camera that provides a wide-angle, constant stare and a targeted moving eye, all in one package. Through extensive interviews, Michel discovered that, in the late 1990s, an unnamed engineer from a US military research facility was inspired by this film to actually create the technology.[42] Years later, multiple teams, one of which was run by McNutt, received millions of dollars of federal funding to develop competing prototypes. McNutt's team won. Though their creation was less powerful than what Hollywood imagined, it came close. Taking their not-quite-perfect system, dubbed Angel Fire, to Fallujah in 2007, McNutt's team was able to successfully track roadside bombers backward in time from the site of an explosion and send teams to "neutralize" the threat.[43]

Once he retired, McNutt thought the technology could be adapted to help domestic police. Rapid prototyping of a mass surveillance technology in a war zone based on an idea from a dystopian Hollywood thriller is a rather uncomfortable fit for the more traditional, procedure-obsessed world of policing, but McNutt saw this as an asset. He sought to bring the same sort of rapid experimentation that had served him well in the military to help disrupt the policing sphere. After several limited trials in other North American cities, he finally got his big shot in Baltimore.[44]

The spy plane first came to Baltimore in 2016 under total secrecy. BPD told no one what they were doing: not the public, not the public defender's office, not the prosecutor's office, not the city council—not even the mayor. Partly, this secrecy was a product of how the program was funded—a private donation from the Laura and John Arnold foundation. The Arnolds are Texas-based billionaires who head up a heavy-hitting philanthropy called Arnold Ventures. John Arnold told me that, in 2015, he happened to be listening to the very same *Radiolab* podcast episode that had attracted me to McNutt's company. Like me, he called McNutt directly and asked if he could find out more about the technology. Impressed by McNutt's pitch, Arnold told him, "If you can find a place to try this out, I'll fund it." While this sounds extreme, it's more common than you might think. Wealthy donors have their fingers in all kinds of technology and criminal justice experiments all over the country.[45]

For Arnold, the spy plane trial in Baltimore was literally supposed to be a field experiment—a test to find out if it really worked. "It's really hard to find causation in the crime space," he told me. "Our foundation looks at a lot of interventions and you just can't say with any definitive conclusion what works and what doesn't." Frustrated by the lack of hard data on the effectiveness of different programs and technologies, Arnold thought he could put his finger on the scales by supercharging a long-shot technology like the spy plane in any community that was willing to try it. McNutt approached officials in Baltimore because of the city's reputation as a "laboratory for spy tech," most notably its massive CCTV network known as CitiWatch.[46] BPD agreed to try it out, especially since it was free to the city and would therefore not have to be disclosed to the public. Two donations from the Arnolds totaling $360,000 were passed through local Baltimore charitable organizations so that the program would not have to be publicly vetted by the city's spending board.[47] The program was thus both secret and free for the police to "check out," all because a billionaire listened to a podcast.

No one was more concerned about the secrecy of the 2016 program than McNutt, though he certainly appreciated the Arnolds' free ticket to experiment. As we shall see, for McNutt, the spy plane is primarily about deterrence. In order to have that effect, though, people need to know it exists. Would-be criminals must fear the plane could be watching. Moreover, public controversy over the technology had halted PSS elsewhere, such as in McNutt's own hometown of Dayton, Ohio, so he wanted to get out ahead of any ill feelings among citizens by having a slow, transparent rollout. In a series of emails, McNutt pleaded with BPD officials that PSS be allowed to hold "focus groups" with residents of West Baltimore prior to launching the plane, in order to get the word out and "gauge community acceptance and concerns." Police refused. They said they wanted to see if the plane worked first, then roll it out publicly.[48] This made no sense to McNutt, he told me, "because the whole way it works is by making it public." Nevertheless, hungry to get a shot to put a plane above one of the most troubled American cities, McNutt went forward with the program anyway.

Before BPD could go public, the existence of the 2016 spy plane trial was revealed by journalist Monte Reel in a splashy long-read for *Bloomberg Businessweek*.[49] Reel leaned heavily into doomer themes in the piece, shredding PSS's reputation.[50] When the article hit, the public firestorm McNutt had predicted, and about which he had warned BPD leadership, broke out. The *Baltimore Sun* editorial board immediately denounced the spy plane and called for its grounding. "Big Data Is Watching," the headline read.[51] Despite having flown only a few hundred hours, the program was quickly shuttered. McNutt expressed deep regret about this failure to me. "We weren't even up long enough to show what we can really do," he told me in 2017. He longed for another shot.

While McNutt made a lot of enemies in the city from the 2016 trial, perhaps his most important ally, John Arnold, remained interested. A moderately positive evaluation of the program in 2017 by the prestigious Police Foundation gave McNutt's company a fighting chance to pursue another trial, if the political winds ever shifted.[52] In part, McNutt *made* the political winds shift. As I explore in this book, his determined effort to bring the program back to the city, which spanned nearly four years of pitching to and convincing residents of West Baltimore to embrace the spy plane, ultimately bore fruit. In December 2019, on the heels of another nearly record-setting year of murders, a coalition of West Baltimore citizens, business leaders, and politicians convinced Baltimore's police commissioner to let the spy planes watch the city once again. Arnold signed on to the program reboot, ponying up $3.7 million for a massive public rollout that the police commissioner cleverly called the Aerial Investigation Research (AIR) pilot program.[53]

What happened next has never been revealed to the public. I tell the story in a strange sequence, much like how spy plane analysts move back and forth in time. In the next chapter, I show an example of the plane "working as advertised" (a rare instance) by helping detectives arrest a clever and evasive homicide suspect. Chapter 3 describes one of the program's most disturbing and unforeseen problems—a "false positive," in which analysts mistakenly tracked the wrong person. These two examples lead to deeper themes. I "rewind" and show how the plane fits within the long history of police experimentation in Baltimore, describing the city's deployment of other controversial anti-crime tech (chapter 4). Moving forward in time again, I describe how the West Baltimore citizens who backed the program, as early as 2017, were convinced of its utopian promise as a way to hold police accountable and restore trust (chapter 5). The final chapters return to 2020 and beyond, discussing the program's legal controversies: how PSS ended up violating the Fourth Amendment (chapter 6), and how the evidence from the spy plane failed in the court system and upended the lives of suspects and victims' families (chapter 7). In the conclusion, I make some recommendations for how we can think beyond the hype to more responsibly deploy new technologies in the criminal legal space in a way that can empower Black and Brown communities.

2   The All-Seeing Eye

Seeing a murder with the spy plane, at least when you get a good case, is like time travel. It's like being handed a mystery novel opened to the middle chapters. You know that the whole story is there in the book, already written down, but you're not sure how you got to the particular page you were given, and you're not sure what happens at the end. So, you have two tasks: one, to go back to an earlier part and work your way through to the moment of the crime; and, two, to push your way forward to the end of the story, hopefully finding out who the killer is along the way. Outside the story, in real life, time marches forward like normal. The case gets older and older, colder and colder, as investigators push forward with what they have or decide to drop the case. But inside the story, everything is frozen in time. As you become more familiar with a case, you can move around more freely in the world of the story—back and forth in time, revisiting key parts, and gleaning even more meaning from earlier bits, because you already know what happens in the end. The murder is always there. Preserved.

Around three in the afternoon one summer day, two detectives are let in through the password-protected door to the spy plane operations center. Occupying half a floor of a large office building in downtown Baltimore, the center is what you might imagine from a television crime drama: filled with gigantic screens and people looking intently at images and paperwork about crime. A few months earlier, when the head of the company, Ross McNutt, was showing me around the place as it was being constructed, he said he wanted detectives to be impressed with what they saw, so he put up as many screens as he could possibly fit in the room. There are an almost comical number of screens.

Gigantic monitors sit in front of each member of the team he hired, consisting of twenty analysts and five supervisors, arranged evenly in four rows of long desks. These fifty-inch screens allow them to look closely at the pixelated images from the spy plane cameras. As the two planes fly above the city in adjacent circles, each equipped with a wide-area motion imagery (WAMI) unit, they capture one massive image of thirty square miles of the city each second. Click. Click. Click. The images aren't perfect. A car looks like a little Tic Tac, and a person like a little dot or smudge, but the resolution is just enough for a sharp-eyed analyst to follow people and cars on the ground.

Using iView, the company's proprietary image processing and markup software, each image is rendered in seconds, mapped onto its corresponding latitude and longitude on the earth, and tagged with a precise time stamp. The result: an almost perfectly preserved visual record of the second-by-second movements of almost the entire city in both time and space. The software allows analysts to zoom in on smaller sections of the city and make tracking marks using colored circles and lines. Again, the zoom isn't perfect. They can't go from a city-size view to looking at your face, but, as shown in figure 4, vehicles that are completely undetectable to the naked eye at the image's widest view become just clear enough to see and tag when you zoom in. The gigantic monitors, then, are key to giving analysts a fighting chance of actually picking out a tiny person walking on the street and following them each time the fuzzy, jumpy imagery refreshes.

FIGURE 4. Two views from the same spy plane footage. *Top:* Zoomed out on West and Central Baltimore. Rectangle is location of image below. *Bottom:* Zoomed in on a street corner. Circles mark two moving vehicles.

Up on the walls, televisions are hung displaying weather, news feeds (Fox News on one side, CNN on the other), and other information from ongoing investigations. These screens are almost never used by the analysts, as far as I can tell, but they give the room a certain Hollywood "situation room" feel. Though it is scorching outside—it's July—the air conditioning is blasting to keep all these devices cool, so everyone is wearing an extra layer.

I recognize Detectives Franks and O'Reilly as they come through the door—two young and ambitious homicide detectives who are part of a squad that has made ample use of the spy plane. Earlier in the day, O'Reilly called in to say that a homicide had happened around 12:30 p.m. on the west side of the city and to get the imagery ready, because they would be coming to see the footage as soon as they were done canvassing the scene. In an unusual move, McNutt had the footage downloaded remotely from the plane while it was still in the air, rather than wait until the end of the day to have it delivered by hand on a hard drive. By the time the detectives arrived, Jessica—the program's head analyst supervisor—had the imagery ready to go.

Jessica has Franks and O'Reilly crowd around her terminal to take a look at the imagery on iView. They tell her the exact time of the murder—12:47 p.m.—and the exact address of the incident—a "car-ryout" convenience store on Carroll Avenue in a residential neighborhood of West Baltimore. The investigation would thus become known as the Carroll Avenue case, since police refer to homicides by the street name rather than the people involved. Jessica uses some software tools to pull the time back on the imagery and zoom in toward the address. "It's in coverage," she tells the detectives. Zooming down some more, she says, "And it's looking really clear. Right in the center."

As illustrated in figure 5, there is a basic procedure to a spy plane investigation, and several questions come up early on that can ruin

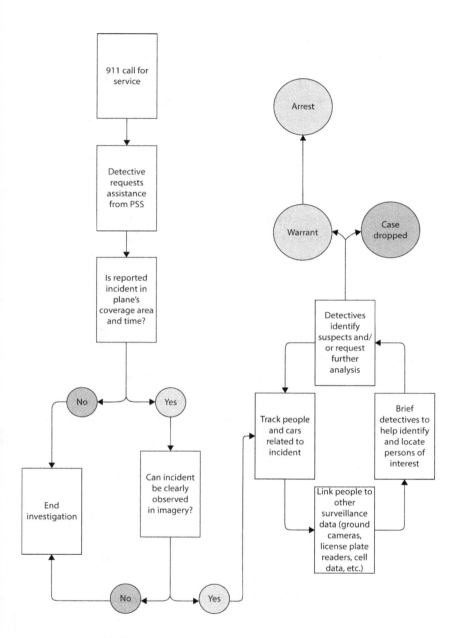

FIGURE 5. How a spy plane investigation unfolds.

the case from the start. In this instance, unlike many of the cases the program received, many variables have lined up to make the Carroll Avenue imagery particularly good. The first is that the plane was actually in the sky during the incident. It is often down because of rain, low clouds, or darkness. Beyond this basic issue, there are also many factors that contribute to the quality of the imagery, making it suitable to actually see the incident in question and track accurately. In this case, the plane had been in the air all day that day and would continue through to sunset, allowing analysts to go far back and forth in time. Nothing is more frustrating than when the plane captures just a few minutes of a crime before it lands, say, to refuel. The incident is also located close to the center of the plane's orbit where the imagery is the crispest. Cases on the edges of the image can be hard to work because the footage is blurry. And finally, the footage is taken from an altitude of around seven thousand feet. Unlike what you would expect, the lower the plane flies, the worse the imagery is for tracking. When the plane is pushed to low altitudes by cloud cover, for example, the cameras create a more angled view of the ground, rather than a top-down view, which makes tall buildings and trees appear to stick out more. These can easily obstruct an analyst's view as she tracks. A line of trees or tall buildings can appear to lean over and completely obscure an entire road sometimes.

This time, the imagery is perfect. The detectives look at each other approvingly. They have been burned by these kinds of complications before, and given that the murder is only a few hours old, they are able to do more with the plane than just document what happened or open up a few leads. They might actually be able to find out where the suspect is currently located and make a quick arrest. That would certainly "send a message" that police are watching.

Jessica zooms in so that the carryout and its adjacent streets and buildings fill the whole of her monitor. She moves her head closer to the screen. Franks and O'Reilly lean in, too. She starts quickly clicking through the imagery from 12:46 onward, each click of her mouse serving up the next second of recorded motion. At around 12:48 she sees a group of tiny dots moving rapidly in different directions out of what is probably the entrance of the carryout. "There," she says. "Yep," says Detective Franks, "that's it." Here are all the witnesses fleeing. Franks says that they already pulled the security camera footage from inside the carryout and noted that there were about seven or eight people inside who fled the store right after the suspect had walked in and shot someone. Franks and O'Reilly pay almost no attention to the witnesses; I assume this is because they know none of them will talk about what they saw. Nobody trusts the BPD in this part of town.

The detectives tell Jessica that, in the security camera footage, the suspect runs out before the witnesses. Jessica pulls the imagery back in time and again starts moving through it, frame by frame, now with an eye more attuned to where the carryout entrance is. She suddenly notices another dot exit the store very quickly a few seconds before the larger group. She had missed this detail the first time. Pointing at the dot, she remarks, "Whoa, look at this guy." "Yep," Franks agrees, "that's what we saw [in the security camera footage]. A guy runs out."

With Jessica having identified the basic crime scene and some suspicious movement, the spy plane imagery is now able to start revealing more of the story. Jessica points to the little dot that is presumably the suspect fleeing the scene. With her finger a few inches from the screen guiding onlookers' eyes, Jessica helps us watch the little dot of a person move at very high speed to the left, around some sort of obstruction (maybe a fence?), into an alleyway just behind the store, and up to a dark-colored vehicle that's parked in the alley. The dot disappears, one presumes, inside the vehicle. Two or three seconds later the vehicle begins to move at high speed down the alley and through the neighborhood, heading southwest, and then circling back eastward toward a large and busy intersection. Before the car has to stop, with the traffic piling up at the intersection (presumably it's a red light), it makes a series of wild shortcuts southbound through a parking lot, into an adjacent parking lot, and out onto a busy road, doubling back and now heading northbound. "He's avoiding that stoplight," Jessica surmises.

She centers the imagery on the parking lot and clicks on a button on the iView menu called "Google Earth." The software automatically loads a Google satellite image in a new window cued to the exact same location of the parking lot displayed on the iView screen. She zooms down into the Google satellite imagery until it automatically loads a Google Street View image of the parking lot (date stamped June 2019). It reveals that the suspect was driving through the parking lot of a gas station. O'Reilly says, "Oh yeah, we know that place. It has really good cameras, so we can pull that." Franks writes down the exact time recorded on the spy plane footage of when the vehicle goes through the parking lot; that way, when he visits the gas station later, he doesn't have to spend hours looking through the entire day's security camera footage to find the vehicle he needs.

Everyone steps back from the screen. The detectives, looking satisfied, tell Jessica that they will head over to the gas station now and get in touch if they need more. She says her team will get to work on the case.

The detectives leave. Jessica looks at me with raised eyebrows. "That was nice," she says. This is already a good result. Without the plane, the detectives would never have known to look at that specific gas station. And even if they did, they would not have known the exact time period to pull up in the station's video logs. A case that could have easily gone cold is now red hot.

Does Crime Technology "Work"?

The Carroll Avenue case, and indeed the whole spy plane program, raises a host of thorny questions about the risks of giving the police a supposedly all-seeing eye. Most pressing, many would argue, are questions around privacy and how the technology gives the state a kind of time machine, allowing police to rummage through our lives. I take up that issue later, but for now I want to address an often overlooked question about the spy plane and about policing technology more generally: does it work? This is an important question because, even if the technology contributes to an invasive surveillance state, if it manages to reduce violence in a city like Baltimore, my guess is that few will care about those criticisms. Especially for those who live with the daily threat of violence, concerns about something as hard to pin down as privacy can pale in comparison to the worry about one's own life and the lives of family and friends.

While assessing whether a surveillance technology "works" sounds easy, it turns out to be a very tricky question. One way we might look at it is to ask whether having the plane in the air can reduce violent crime (and particularly homicide). When law enforcement officials say a surveillance technology is effective against homicide, they usually mean it can reliably aid in the identification of suspects in a way that leads to an arrest and prosecution of the person who actually committed the crime. If it can do this, their thinking goes, it will contribute to a reduction in violence through a "deterrence effect"—sending a warning to citizens that those who hurt others will be held accountable . . . so just don't do the crime. As McNutt likes to say, he hopes his technology never has to solve crimes because it has deterred them in the first place. Deterrence, criminologists have found, seems to arise when there is a widespread sense of certainty that one will get caught if one breaks the law. It's not about severity or style of punishment; it's about quickly and reliably getting seen, identified, and arrested.[1]

To understand how the spy plane might have a deterrence effect, you have to know something about why detectives have such a hard time identifying, locating, arresting, charging, and verifying the guilt of a suspect in a homicide in the first place (called "solving" a case). City police across the United States, including Baltimore, solve less than half of murders and shootings, sometimes barely scratching 30 percent.[2] One of the biggest barriers to solving cases, it turns out, is a lack of human witnesses. In Baltimore, as in many other high-crime cities in America, the people with the most important knowledge of a suspect's identity and whereabouts often refuse to talk to police. This is true even when they are the victim's family members. Baltimore is notoriously the place where the phrase "stop snitchin'," the tagline for a culture of witness silence, was popularized.[3] BPD detectives constantly referenced this problem during my fieldwork and suggested that their jobs would be orders of magnitude easier if people would just tell them what they saw.

Part of the reason for this code of silence is that citizens tend not to trust Baltimore police because of a history of aggressive, racially biased, and unconstitutional zero-tolerance policing—a history that has been well documented by a recent Department of Justice investigation.[4]

Additionally, citizens are afraid of retributive violence if they snitch to police. As the saying goes, "Snitches get stitches."[5] They do not trust that police can protect them if they offer information. In the summer of 2020, for example, Instagram revealed that an account had been made that published faces and court documents outing people who had talked to investigators in Baltimore, presumably to scare them out of continuing to cooperate. After Instagram took the page down, another one immediately appeared.[6] Who would risk such exposure to give information to a detective they already don't trust and who can't protect them if they speak out?[7] The trust crisis makes detectives' jobs a lot harder.

One way of thinking about whether the spy plane works, then, is to ask if it can intervene in this crisis of trust and witnessing, not by mending that trust but by finding a technological work-around in the form of a panoptic system—an all-seeing eye in the sky. As McNutt often claims, the promise of WAMI is that it can be an "independent witness" to violent crimes with near-perfect vision and a capacious and infallible memory, which could relieve community members of ever having to snitch. That claim rests on two assumptions about how the plane works: first, that it can archive the "ground truth," as McNutt puts it, of a crime scene—a visual record, stored on a secure server, of what happened at the scene that is not biased by any particular point of view, failure of memory, or partiality; and second, that it can tell who was there by linking individuals at the crime scene to other surveillance data that can make an objective identification (usually by tracking a person to CCTV or security camera imagery outside the crime scene). If the plane can do these two things, the thinking goes, it can get around the problem of witnessing, which would be an immediate relief to both investigators and citizens. It would also make PSS a hugely sought-after company and McNutt a very wealthy man.

The Carroll Avenue case, as I will show, is probably the best example of the spy plane "working" as advertised. If that's the case, then, how did it work?

Virtual Stakeout

After the detectives leave, Jessica spends a few more minutes going back through the imagery again to "lay down track points." Starting at the few seconds prior to the suspect fleeing, she uses iView to put a small colored circle directly over the little dot of a person as he runs. Frame by frame, she clicks right on top of the suspect's pixelated body. Strung together, these circles form what analysts call a "track." She labels the track "Subject 1." By laying down points, she is making this an official track. Each track point, with its time and latitude and longitude coordinates, is linked to her ID card and terminal.

Whatever happens from here on out could theoretically go into evidence with the entirety of detectives' work on the Carroll Avenue case.

Having officially opened the investigation, Jessica then assigns the case to one of her three teams of analysts. Over the months, Team 3 has become the go-to group for homicide investigations. Composed mainly of young Black men and women who grew up in West Baltimore, one of the neighborhoods most deeply affected by gun violence, the team has proven itself superior at handling the most critical and complex cases covering portions of the city that they know intimately. Most of them are no strangers to Baltimore's violence. It has not been uncommon for them to conduct surveillance right over their own houses.

Because this is such a time-sensitive case, it is important that the tracking be quick. Jessica therefore divides the tracking in two. Ejaife, a thirty-something Nigerian American who always comes to the office impeccably dressed, takes the "post-incident" track—following the suspect's path forward in time away from the crime scene. His goal is to get the suspect going through as many other surveillance sensors as he can find, usually CCTV cameras, and ultimately to a final stopping location where the suspect might be intercepted by police. She gives Mark, a tall and lanky twenty-something with an infectious smile and quiet demeanor, the "pre-incident" track. His job is to track the suspect backward in time from the crime scene to his origin point, again trying to link the target with other surveillance sensors.

The rest of the team begins familiarizing themselves with the layout of the public CCTV cameras in the area. As part of their agreement with the city of Baltimore, spy plane analysts were granted direct access to CitiWatch—the BPD's vast network of publicly owned cameras—right on the company's private data terminals. They can pull up any camera in the city, punch in the date and time, and start watching the archived footage.[8]

CitiWatch Ground Camera Integration
White Infiniti In Camera Views

Camera 715: W North Ave – N Payson St
Suspect 1: White Infiniti 10:14:51

Camera 715 W North Ave – N Payson St
Suspect 1: White Infiniti 12:02:23

CSP Vehicle tracks with CitiWatch In Camera Views
CitiWatch Cameras

Time	Level	Comment
10:02:32	Low	929: Edmonson Ave and Allendale St.
10:05:09	Low	821 Mt Holly Street
10:14:52	Low	715: W North Ave and N Payson St
10:50:14	Low	722: N Monroe St and Wallbrook Ave
10:50:28	Low	718: N Payson St and Ridgehill Ave
12:02:12	Low	721: W North Ave and N Monroe St
12:02:23	Low	715: W North Ave and N Payson St
12:05:52	Low	721: W North Ave and N Monroe St
12:06:11	Low	712: W North Ave and N Pulaski St
12:06:20	Low	709: W North Ave and N Smallwood St
12:06:26	Low	737: W North Ave and Bentalou St
12:06:41	Low	706: W North Ave and Moreland Ave
12:07:38	Low	905: W North Ave and N Dukeland St
12:08:28	Low	910: Baker St and N Dukeland St
12:09:40	Low	916: Bloomingdale Rd and Ellicott Drwy
12:10:39	Low	918: Poplar Grove and Riggs Ave
12:11:07	Low	920: Poplar Grove St and W Lafayette Ave
12:11:34	Low	921: N Franklintown Rd and N Longwood St.

FOUO Law Enforcement Sensitive

FIGURE 6. Using a tracking report to pull up corresponding CCTV footage of a target vehicle. (*Source:* Persistent Surveillance Systems. Used with permission.)

The company's software engineers also integrated the locations of these cameras into the aerial surveillance imagery as a digital overlay on iView. With the click of a button, analysts can lay the locations of every CitiWatch camera in the city over the wide-angle footage. Each camera appears on the spy plane footage as a small white ring with the camera number and approximate address. Much of what analysts do, as seen in figure 6, is use corresponding time stamps in the two systems to link observations from the iView imagery to ground-based CitiWatch imagery, allowing them to see both broadly and closely at the same time.

Ejaife picks up the tracking where Jessica left off and immediately gets the suspect going through three CitiWatch cameras. He passes the camera numbers off to his teammates with, "Hey, somebody check 154, 155, and 183." Bina, sitting next to Ejaife, pulls up each camera feed and sees that they are all broken; nothing but blackness. "They're all down," she reports. "Every. Fucking. Time," Ejaife grunts. This has been a continual problem. It seems like whenever analysts need one of the city's ground cameras the most, it fails them. Most of Baltimore's CCTV cameras "pan and scan," rotating back and forth along an intersection or road, rather than pointing in a fixed direction. A running joke among analysts is that any time a person gets shot in Baltimore, the CCTV camera makes sure it is panning away, just so that it *won't* capture the crime. It is also not uncommon for city cameras to have malfunctioning scanning patterns that point at a wall, inside somebody's window in a public housing complex, or upward toward the sky. Or it just doesn't work at all and nobody has come to fix it yet. Every single camera blank, though? This is unusual, even for Baltimore.

Suddenly, thunder booms outside and everyone realizes that a huge summer storm is coming through. McNutt, who has been alerted to the case and has come in from another part of the analysis center, tells his team that the whole CitiWatch system is down temporarily because of the storm. The archived footage is still there; there's just no way to access it right now. The analysts will have to go on nothing but iView for the time being.

Ejaife keeps tracking the suspect's vehicle forward and sees it going down a narrow road in a residential area. He sees him back the car carefully into a little tree-covered spot at the end of the street. Because of the tree cover, it is impossible to see from the aerial footage if anyone gets in or out of the vehicle. Ejaife can only *just* tell that the car is parked there because the tone of the tree cover is made ever-so-slightly darker by the car sitting under it. Only a trained eye would notice this. To most of us, it just looks like a blob of shifting gray splotches. Ejaife has to keep going back and forth through the imagery to make sure his eyes aren't tricking him. He is confident that the vehicle is actually there and hasn't moved. He pulls up the corresponding Google Earth imagery of the street and sees that the suspect has parked in an alleyway between two abandoned row houses—probably not where the suspect lives, but certainly a good place to hide a vehicle, the analysts speculate. Clicking forward in time, Ejaife sees that the car sits there for several minutes. He skips the footage ahead another ten minutes and sees that the car is still in that spot. He skips it forward all the way to the end of the footage and, again, the car is just sitting there. Is this the end location of the suspect's path? Given that the footage ends at 3 p.m. (the plane had to come down early because of the storm), and it is currently 5:30 p.m., there is a good chance that the car is still there at this very moment.

Jessica immediately gets on the phone with Detective O'Reilly to relay this important news. He says they will head out now and see if the vehicle is still there. The whole analysis center gets a little quiet at this point. This is a big deal. Based solely on the spy plane's recommendation, the police are about to engage a suspect on the ground. Imagine if they got the tracking wrong, sending the police to a person with no involvement?

A call comes back just a few minutes later from Detective Franks. They went to the address. The vehicle is still there and they are confident it's the same one detectives saw in the CCTV footage they pulled earlier: a black, two-door Honda Civic with tinted windows. Having totally lost the car earlier that day, the detectives are now looking right at it. McNutt and Ejaife jump out of their seats and high five. Everybody in the room starts clapping and Ejaife is beaming. He pumps his arms, yelling, "I helped Baltimore today!" Jessica looks at me, smiling, and says, "This is the way it's supposed to be."

It has not always been like this. These analysts have become more used to working on days- or even months-old cases for detectives who have already tried everything else and gotten nowhere. These stale cases have been almost worthless to investigate. Any retrospective information that the spy plane has is usually sorely out of date. Oftentimes detectives have moved on to the next homicide and don't even look at the evidence packets that are delivered to them after hours of painstaking work. This time was different. The detectives came over to see what the imagery looked like right away; they acted on the intel immediately, fed intel back, and that led to more leads. And now they likely know where, if not who, the suspect is at this exact moment.

Jessica gets another call from Franks. They now have eyes on the car and are trying to decide if they should tow it or just sit on it to see if the suspect returns. Jessica tells him that, as far as they can tell, the car hasn't moved all day, but they will keep looking. Now everyone in the analysis center has dropped what they were doing and hopped on this case. McNutt tells everyone to take a look at the car's final location and the incident location to look for anything else suspicious. All the analysts silently follow different cars and people in and out of these areas, just "eyeball" tracking and not laying down track points. It is dead silent. McNutt suddenly shouts, with a grin, "Virtual stakeout!" Everyone chuckles.

Counting Cars

My attention now turns to the pre-incident track that Mark has been working on for the better part of an hour. He has followed the suspect backward in time—northward from the incident location for quite a ways. Confusingly, because Mark is tracking the car northward in backward time on iView, this actually means the suspect was heading south in real, forward time. Over the months, analysts have become better and better at understanding this reversed temporal world—arriving is actually departing, stopping is starting, a left turn is actually a right turn, and northbound is southbound. Whereas they used to have to pause and think about this, manipulating the direction of time in their minds, now it is like second nature.

There are a lot of tricky spots in this section of tracking, and Mark is having trouble. In one part of densely tree-lined road, the car keeps disappearing for several frames behind overhead obstructions—a very common issue in this old city of tall trees, narrow streets, and three-story row houses. The car pops out for one frame between two trees and then goes hidden again as it moves along. In these situations, it is remarkably easy to accidentally "jump cars"—go from tracking the correct vehicle to the one just in front or behind—without realizing you've made the switch. Mark has to use a technique analysts call "counting cars." During training, analysts learn that just because a vehicle is hidden for a few frames doesn't mean you can't figure out where it went. It just takes some logical deduction. Pulling the imagery back to the moment where the black car enters the foliaged area, Mark counts the number of lighter colored and darker colored cars both in front and behind "his" car (the suspect's car). He then pushes the imagery forward again and begins watching the area where the foliage stops, counting each light and dark car as it emerges. Assuming the cars haven't radically changed speed or direction, the order of cars should be the same before the foliaged area as after it. "There he is," Mark murmurs. He continues to track for a dozen frames or so and then stops and pulls his teammate Adam over and says, "Hey, make sure I didn't jump. Is this my car?" Adam takes a look, replaying the imagery several times. He turns off Mark's tracks so he is not biased by his teammate's analysis. "Nope," he says. "Yeah, see you jumped right there." He points to a dark car running one car in front of the one Mark tagged. "That's your car," he says. "Are you sure?" Mark asks. He goes through it again and says, "Damn, you're right."

Mark proceeds with the track on the correct car, assured that he has fixed the mistake. (Later, Mark tags the car he mistakenly tracked for a dozen or so frames "Determined Not Involved," in order to document the error.) He watches the car weave its way into a residential neighborhood and park (in actuality, it was departing from a location and weaving its way out of the neighborhood toward the soon-to-be murder scene). He uses Google Earth to pull the address and suddenly realizes that it is a house just a block north and east of the final location Ejaife discovered just a few moments ago. This opens up more details of the murder's narrative. It appears that the suspect dumped the car in a familiar neighborhood, an area he visited just half an hour before the murder. Jessica calls Franks and explains the new development, adding, "In case that makes any difference for you to sit on the car or tow it." She hangs up and gives me a knowing look. This information will probably push the detectives to watch the car rather than tow it, given that the suspect likely lives in the area. If they towed the car and the suspect saw them, it might spook him and cause him to flee capture. She widens her eyes and twists her face, the corners of her mouth curling down, as if to say, "I hope that was the right call."

It is now well past the end of everyone's shift and the analysts are burning up overtime. Jessica sees that they are probably not going to get anything more out of this case until CitiWatch is fixed, so she decides to pull the plug. As the team packs up, I notice that McNutt is still looking intently at the incident location. Jessica and I walk over to see what he is doing. McNutt looks at Jessica sheepishly and points to a pixelated person walking down the street near the carryout just prior to the murder. "Can I track him?" he asks. Jessica laughs painfully. Over the months, McNutt has had a tendency to work on cases after analysts have gone home, making dozens of new tracks. When analysts return in the morning, the case is sometimes a mess. No one knows what's going on, why things were tracked, and what tasks are yet to be finished. But like a kid in a toy store, he cannot keep his hands off this shiny new case. Jessica scolds him playfully, "Okay, but whatever you do tonight, please don't make me come back tomorrow and get mad at you, mister." McNutt says he promises to document everything.

Facial Recognition

The next morning, Team 3 eagerly jumps back into the Carroll Avenue case. McNutt kept his promise, and the tracking looks much the same as it did last night, with only one new track—an individual on foot at the crime scene prior to the murder, who may or may not be involved.

Jessica bursts out of the supervisor's office with a "you're not going to believe this" look on her face. She shows Team 3 her cell phone, which displays a photo of a patch of concrete surrounded by forest. It looks like a dead-end street in the middle of nowhere. On the concrete is a big black smudge with ashes and burned and broken bits of metal and glass. "What's that?" I ask her. "That's the car," she says. "They burned it. It's gone." Everyone is shocked. According to the sergeant who heads up Franks and O'Reilly's squad, Jessica explains, the detectives decided to keep surveillance on the car last night, instead of tow it. To maintain surveillance at night, though, the spy plane is of no help because, as part of the agreement with the BPD, it is only allowed to fly during the day. Though the company has a camera unit with infrared capabilities for its international military contracts, they have been wary of using it domestically in the United States for fear of pushback from "doomer" privacy advocates. Spy plane "boomers," by contrast, found it baffling that a technology designed to solve homicides was not allowed to operate at night, when most homicides happen.

Without the spy plane to watch the location, detectives decided to bring in a group of officers from a stolen vehicle task force to keep eyes on the car overnight. Around 11 p.m., however, the officers ran out of overtime hours and had to cut their work short. Before they left the car, then, they decided to put a GPS tracker on it, which would alert them if the car ever moved. Through some sort of countersurveillance, the suspect must have caught wind of the officers' presence. As soon as they left, he went to move the car. But . . . the GPS tracker didn't work. It didn't ping the task force until the next morning. Where did the car end up? In Leakin Park, burned to the ground. This is a disastrous development for the case. Adam looks at me and says, "We should have told them to tow it. Damn!" Though detectives found the likely whereabouts of the suspect, actually having the car in their possession could have yielded important evidence—DNA, clothing, a weapon, and who knows what else. The suspect is in the wind again.

The situation is actually even worse, according to Franks. They were able to recover the VIN plate and license plate from the burned car. Tracking down these numbers led to an interesting insight. The VIN belongs to a car that was stolen from a used car lot way over on the east side of the city two weeks earlier. The plate belongs to a completely different vehicle, which was also stolen from the east side. This suspect is clearly a pro. He stole the car that he planned to use to commit the crime, and then even had the presence of mind to replace its plates with different, stolen plates. Not only did he burn the evidence, but even the pieces that survived are throwing police off the scent.

The pressure is now on for the analysts to come up with a new lead. McNutt hops up and starts telling every analyst in the room to pull up imagery from the past few days, even weeks if they have to, and start looking at the location where the car was stashed behind the abandoned row houses. Perhaps they can see something around the neighborhood that would yield information about who lives there. "Track any black car in the area," he shouts. Team 2 takes the imagery from the day before the murder, and Team 1 takes the day before that. Their task is to just aimlessly watch the entire neighborhood and see if they can find, like a needle in a haystack, the suspect's car and somehow try to make an identification. "Okay, track anything that comes out of that street," he tells one group of analysts. Jessica stops him, fast. "No, we can't do that, Ross," she says. "Why not?" McNutt asks. "It's a privacy thing," she replies. "We have to be careful with stuff like that." "Well, can we at least look at all the cars but not track them?" he pleads. "Yeah, that's okay," she replies. "Just don't put track points down on everything. But it's okay to just visually follow cars in and out of that area."

This distinction, between "eyeball tracking" and "real" or "official" tracking, was a hot topic of conversation in the operations center. When a target is tracked, however briefly, it becomes part of an investigation, subjecting the person to a kind of digital stop and frisk and thus invading their privacy in some distant, but real, way.[9] As I will explore later, the lack of clarity about who can and cannot be tracked would become a key reason why the spy plane was later deemed unconstitutional. At this point, however, those boundaries were still being explored. Whereas McNutt took the position that the program has a remit to track "anything that moves," so long as it can be linked to an investigation, Jessica was much more reserved. Over the months, largely because of her caution, analysts became less and less willing to lay down track points on a target unless they were certain the target was actually involved as a suspect or witness. Still, analysts would routinely track using just their eyes (and often a pointer finger) to see if something was, indeed, "worth tracking." This eyeball surveillance never makes it into evidence.

Soon, Jessica comes over to Team 3 to announce that Rania, from Team 2, just found something. The fact that *she* found it is already significant. A young, soft-spoken Russian American woman, Rania has become known as the most sharp-eyed tracker in the group. Demonstrating time and again a unique ability to see things that no one else can see, she has routinely dug other analysts out of jams when they lost a target. Earlier, Rania had been triple-checking Mark's backward track from yesterday, given his worries about jumping cars. Having confirmed his work, she then "eyeball tracked" the car further back to the beginning of the day and discovered that it had stopped at a gas station about forty-five minutes prior to the murder. The suspect appears to get out of the vehicle, perhaps to fuel up. Jessica asks Bina, on Team 3, to double-check Rania's work. Bina begins to lay down actual track points and confirms Rania's eyeball tracking. Jessica quickly texts the address of the gas station to Detective Franks, who replies that they will head over immediately to see if they can pull camera footage.

Just a few minutes later, Jessica gets an update from Franks. They were able to pull video from a camera on an air pump that the suspect used to fill up his tires prior to the murder. And what did they see? A high-resolution, close-up facial image of the suspect. He texts Jessica a picture from his phone of a computer screen at the gas station showing the suspect from the torso up, standing in front of the pump: a young Black man in a white T-shirt and black ball cap turned backward. The image has a time stamp that almost perfectly matches the time from the iView track. Team 3 high-fives each other and looks thankfully across the room to Rania. This is a perfect result. The analysts pointed detectives to a facial image of the suspect they would never have known how to find otherwise and have now made this case go hot, yet again.

A few hours later, Sergeant Charles visits to take a look at the analysts' work. As the head of the squad in which Franks and O'Reilly operate, he is an important guest. It means this case, and thus the spy plane program, is getting the attention of those higher in the BPD pecking order. Charles tells us that he sent the facial photo from the gas station to an FBI homicide task force that routinely helps out local police. They have facial recognition software. Running the photo against a database of mug shots, the algorithm produced a hit that identified the suspect, who was known to the squad already as a "trigger puller" in the city. Charles thinks they will be able to apprehend the suspect soon.

Just then, Charles gets a call from Franks, who says the suspect was just seen on foot and officers are giving chase. Charles says they are working on a warrant. An arrest seems imminent.

Two Shirts

The next morning—the third day of the investigation—Sergeant Charles returns to work out a few final details about the tracking. They arrested the suspect last night, and after an interrogation, have some new questions for PSS. He pulls McNutt, several supervisors, and all of Team 3 into the briefing room, a small soundproof room used for presentations. McNutt pulls up the footage and begins to show Charles the backward track—where analysts followed the suspect backward in time to the alleyway. Charles stops him almost immediately and asks, "Is there anyone else in that alley?" O'Reilly, the lead detective on the case, pipes up, saying that, during the interrogation, the suspect admitted to being in the alley inside the stolen vehicle behind the carryout just prior to the time of the murder. But when asked if he exited the vehicle and committed the murder, he clammed up. They showed him the CCTV images from inside the store of a man wearing a black shirt committing the crime. "Is this you?" they asked him. O'Reilly says the suspect's story suddenly began to get complicated. He said that another person, a homeless guy wearing a black shirt, approached him in the alley while he was sitting in his car and then went around to the front of the store. That guy must have been the one who did it. O'Reilly wants to know if the plane can shed some light on this story.

McNutt zooms in close on the alleyway and begins clicking through each frame, slowly. With all the analysts and detectives looking along, no one sees anything but one dot of a person getting out of the car and heading toward the murder scene. The alley is empty of people and other cars prior to, during, and after the murder. Charles nods approvingly and remarks, "That's great, that's great." He stresses that the more evidence they can give to confirm that there was only one suspect and one car, the less complicated a story they will need to present to prosecutors.

O'Reilly then adds another detail. In the interrogation, the suspect claims that he was wearing a white shirt that day, not a black shirt (a fact that can be confirmed by the close-up photo from the air pump), and so he could not have been the one in the CCTV footage of the homicide. O'Reilly asks if there is any way to check that the suspect changed shirts right before the murder. McNutt begins to walk through every time the suspect's car stopped. In each instance, ground footage was present to confirm that the same person got in the car that got out and that no other people got in, or even approached, the car. And crucially, the corresponding ground footage before the incident shows a man in a white shirt, and the footage afterward, a man in a black shirt, but both figures are wearing a similarly colored hat. The continuity of the aerial surveillance track, and the opportunity to make multiple visual observations of the suspect, blows a hole in the suspect's "two people" story. It is the icing on the cake of the detective's case: one car, one person, two shirts. They got him.

Everyone busts out of the meeting room in excited chatter, pulling the attention of everyone else in the operations center. Sergeant Charles walks to the front of the room to make an announcement. Standing on a raised platform that overlooks the rows of analyst terminals, and with all the analysts gathered around, he says that they found and arrested the suspect last night. Applause erupts across the analysis center. During the interrogation, Charles explains, the suspect "confessed to almost everything but the shooting." He says that when they went to secure the warrant for the arrest from the prosecutor's office,

> The [prosecutor] was kind of locked up on us. "How do we know that nobody else was in the car?" Well, it was everything you guys gave us. It was the aerial surveillance that you guys showed, that showed a dot getting out of a black car, going to the crime scene, then going back out and getting into a black car, and nobody else getting into the car prior to—that was absolutely critical. So, thank you guys so much. We wouldn't have any of this without you. We wouldn't have facial recognition, we wouldn't have anything because we wouldn't have any of this video. So we can't thank you enough.

Team 3 is absolutely beaming. Whereas the majority of the summer has been spent fighting to get detectives to even pay attention to the existence of the program, this time the spy plane delivered the case on a platter and detectives devoured it.

When I spoke to Sergeant Charles later that day, he stressed again how much the quickness of the analysts' work broke open the case. He said,

> With the way video is nowadays, a lot of stores will have video, but it's only good for like three or four days. If you don't get that video as soon as you can get it, you'll lose it because it tapes over. So, I can't imagine how much video we would have lost had [PSS] not told us how to get back to that video. . . . Literally it saved our entire investigation. . . . You know it's kind of like that movie *Enemy of the State*. It feels like that. It allows us to go back in time and see what happened. And the best part is you're allowed to go back in time and review any portion of the city [*chuckles*]. That's the best part. . . . It's almost like you could go back and replay life for a period of time.

In a moment of life imitating art imitating life, Charles references the very Hollywood film that inspired military engineers to create WAMI in the first place, way back in the early 2000s. Here he is two decades later noticing the similarity as though it was a happy accident.

Technology Is Not Magic

If all you knew about the spy plane program was the Carroll Avenue case, it would seem that, yes, the technology "works." It pointed police to evidence far removed from the crime scene that they would never have known to look for, and it did so in a way that accelerated the pace of the investigation, ultimately leading to a timely identification and arrest of the suspect. Moreover, this was not an easy murder to solve. Though there were multiple witnesses at the scene, none would talk to police. The suspect was clever, was experienced, and used techniques to hide his trail that, one can imagine, must have worked well in the past. And, still, the spy plane found him. When the case came to the prosecutor's desk, moreover, the spy plane was able to alleviate doubts enough to move forward with a charge of murder. While it is nearly impossible to assess empirically, one can imagine a series of arrests like this sending a message of deterrence to some of Baltimore's "violent repeat offenders."

In many ways, then, the Carroll Avenue case is the perfect case for a small, hungry tech start-up like PSS—showing clearly how the technology functions well when it is firing on all cylinders. Throughout its past dealings with other cities in the United States and abroad, PSS has used airtight cases like this one to promote WAMI's almost magical ability to, as the company tagline goes, "solve otherwise unsolvable crimes." When it operated in Ciudad Juárez in 2009, for example, PSS helped solve two high-profile, "witness-less" cartel murders that exposed some of the high-level players in the city. McNutt has taken footage and knowledge from these two cases on the road for years, using them to promote the effectiveness of WAMI to police executives and community groups, including in Baltimore. They form the backbone of the company's analyst training program. You can even watch them on YouTube.[10] Like these showpiece cases, from the point of view of presenting how, in one particular instance, PSS can magically catch murderers, the Carroll Avenue case illustrates WAMI's potential to have an impact on violent crime in a city as desperate as Baltimore. It will almost certainly become a major part of PSS's promotional materials in the future.

From another perspective, the Carroll Avenue case shows the exact opposite. All the reasons this case was exceptional, exciting, and, to put it in Jessica's words, "the way it's supposed to be," suggest that it is a rare instance that the plane would work so well. A study by the RAND Corporation, for example, showed mixed results in terms of case closures. Depending on how you slice the data, some models showed the spy plane delivering better case closures than investigations without it, some models didn't.[11] I am not surprised that the effectiveness study was inconclusive. I am continually struck by the sheer number of factors that must perfectly line up for the spy plane to meaningfully affect a case, let alone contribute to a widespread deterrence effect by *reliably* closing multiple cases. To solve even one case, things like weather, ground cover, imagery quality, ground camera positioning, the speed and quality of detectives' response, accuracy of tracking, and a political climate that would take a risk on an untested technology, among many other factors, all have to be unified in the same cause. And that does not even consider the fact that, as I explore in later chapters, at the end of the case's journey through the criminal legal system, a judge or jury might look at this pixelated imagery and simply not believe that the "dots" in the footage are the same as the people and cars the police say they are. In direct tension with those who champion the plane as an objective witness, the imagery is anything but clear to the average eye. A lot of faith must be placed in the experts' interpretation of it. This is true even on the floor of the spy plane operations center, where multiple analysts are needed to evaluate and double- or even triple-check the accuracy of tracking, and where worries about accidentally tracking the wrong person are always in the background.

So, we're back to where we started: the problem of witnessing and trust—the very problem for which the program was supposed to be a technological workaround. The spy plane intervenes in the trust crisis surrounding Baltimore policing by moving the discretionary power to decide who is and is not "suspicious" up a level of abstraction to a private, for-profit technology company, watching from the air. As sociologist Sarah Brayne observes, a new high-tech policing tool "does not *replace* discretion, but rather *displaces* discretionary power to earlier, less visible (and less accountable) parts of the policing process."[12] But what makes the discretion of a private tech company any more trustworthy than that of police? A lack of trust in the company's reputation, procedures, or any number of other factors I mentioned could create complexities that keep the plane from contributing to a case, or even sink the entire case. As I discuss in the next several chapters, more often than not, this is exactly what happened. Most cases I saw were not this good, and even the good ones ran into significant roadblocks because of mistrust.

Let me go just a little deeper. What do we even mean when we ask if the spy plane, or any police surveillance technology, "works"? Behind this question is an assumption that technologies work mechanically, like a light switch. The question of effectiveness constructs a police department as a neutral canvas upon which is painted "technology," as though the quality of the canvas and shape of the frame have nothing to do with the look of the picture in the end. The plane is put in the sky, and *it* sees guilt or innocence. The camera is attached to the officer's chest, and *it* sees misconduct. The algorithm scans a face, and *it* reveals a person's identity. With all these tools, the question often becomes about how well the coding, cameras, and circuitry mechanically carry out these operations. What my experiences with the spy plane revealed to me is that there are always humans and human organizations in the loop. Technology *is* people. Scholars of technology, in fact, have even created an entire concept to articulate this point: the "socio-technical system."[13] All technologies are a complex intertwining of human organizations and machines. As much as we sometimes wish technology would just "work" for us, there are always humans making and pushing the buttons, and interpreting the data, who are working within organizations constructed by humans that incentivize certain behaviors. The plane doesn't see crime; the analysts do. Even when they see it, much still depends on detectives, prosecutors, journalists, and other humans to pass their interpretations forward, like a baton in a relay race, toward an outcome that other humans label "effective." Whether or not a police surveillance technology works is thus not a vastly different issue from whether or not the entire criminal legal system works. What strikes me is that by relentlessly asking about the effectiveness of this or that technology, we subtly disengage from the deeper question of whether the criminal legal system is, itself, fit for purpose. More worrisome still, we pin our hopes on a magic bullet.

This is the big lesson: simply putting the plane in the sky does not automatically deliver insight into the "ground truth" of crime or pave the way for a justice that might—*might*—give potential criminals pause. It takes a lot of work, a lot of luck, and faith in people. There is nothing magical about it.

In the next several chapters, I begin to peel back the layers of the spy plane program—eventually looking at the entire infrastructure of Baltimore's surveillance apparatus—revealing just how much the city's supposedly high-tech approach to policing rests on hype and performances. Trust, communication, perception, interpretation, bias, and political rivalry all play a part in producing surveillance evidence. More often than not, I discovered, these frailties conspired to prevent the spy plane from ever becoming something as powerful as Big Brother, but also generated new risks and harms that, almost invariably, Baltimore's Black communities were forced to shoulder.

3   False Positives

There was a funny phrase that briefly went around the spy plane operations center: "the Pulaski Effect," it was called, or sometimes "getting Pulaskied," or even "pulling a Pulaski." It was a way of talking about some of the bizarre visual tricks that one runs into while trying to track fuzzy pixels and blurry blobs. "The more certain you are, the less right you are" is one way I heard the Pulaski Effect defined. It is a way of describing a "false positive" in spy plane analysis—when an analyst mistakenly tracks a "suspicious" person or car that actually has nothing to do with a crime. Pulaski refers to a highway in East Baltimore where PSS had one of its first armed carjacking cases in May 2020. It is also where supervisors made a big mistake: they tracked the wrong cars. Even more worrisome, the only reason the mistake was caught is because supervisors assigned the imagery to some new analysts as a training exercise and *they* noticed the mistake.

How is this possible? How could analysts be so certain about what they had seen, but be so wrong? Isn't the spy plane supposed to produce a "ground truth" about crime scenes? After hearing about this case, I had to see for myself. One day, during one of the many lulls in active investigations, I took out my key card, plugged it into a nearby terminal, and pulled up the Pulaski case. The request from detectives was incredibly vague, rife with typos, and short on punctuation:

"Victim vegicle 2010, Black, Toyota Camery,. . . . Suspect Vehicle 2015, white, Acura. . . . Victims vehicle was struck by the suspects vehicle, when they exited there cars the suspect robbed the victim." The time and location of the incident also looked anything but solid—1540 hours on a generic block number of a long stretch of highway. "That's it?" I thought. As I'm taking a first glance at the imagery, Zach, a supervisor who had done the initial tracking on the case, comes over. I tell him that the request looks really unclear. He says he thought the same thing and called the detectives right away for more details. All they could add was that the two cars were heading eastbound when they collided on the highway and one guy robbed the other guy. After the robbery, the victim (in a black car) drove off before the suspect (in a white car), they told him.

Turning to the raw footage, Zach helps me navigate in time and space to look at two cars. Just as detectives had described, one is light colored and the other dark, they are in the approximate time range given, and they are heading eastbound. I start to click through the footage and see the cars come together quite closely, possibly hitting each other. Both cars immediately pull off the highway into a parking lot. Zach points out how there are a few flashes of pixels in the space between the cars that, he thought, could be the robbery. The dark-colored car drives off, followed by the light-colored car. This matched the detective's details perfectly. They put together a quick briefing and sent it off. Case closed.

Zach tells me that a few hours later he noticed two new analysts sitting around with nothing to do, so he assigned the imagery as a training exercise. It was so simple and straightforward—a perfect learning opportunity. One of the trainees was Rania, whom everyone would later learn has some kind of superhuman ability to interpret pixelated imagery, but who at this time was just a newbie. After sitting with the footage, she approached Zach at the front of the ops center. "I think you guys have the wrong cars," Zach recalls her saying. Pulling Zach to her screen, she showed him another pair of vehicles that fit the location and time frame, also light and dark colored, but heading *westbound* on Pulaski. Zach points me to these vehicles, which sit literally on the other side of the street from the original two that were tracked. The two cars clearly come together—no question it's a fender bender—then stop right in the middle of the highway at odd angles, cars swerving wildly around them. The vehicles stay there for a few minutes—certainly enough time for a robbery to occur, but you can't actually see anything. Then the suspect car pulls away *before* the victim's, again the opposite of what detectives said. Zach called the detective to ask about the confusion. The detective confirmed that he had told them the wrong information. He must have gotten it mixed up.

What are the chances that two sets of cars, in the same time and place, would move in such a way that they could both fit the basic outline of this case? There was nothing seriously unusual about that first pair of cars. The second pair were clearly much more "suspicious." Yet, detectives' mistaken information had set off a cascade of mistakes, causing analysts to look right past the real crime. It could have easily gone completely undetected had it not been given a second look as an unplanned training exercise. The more certain you are, the less right you are—that's pulling a Pulaski.

It turns out the Pulaski Effect—the risk of making false positives—was more common than anyone in the operations center had thought possible. Though the technical power of the camera system to create a database of public movements was obvious and ironclad, the investigative power of the analytical team to *interpret* that database accurately during a live deployment was less firm. The data might be objective, but the humans interpreting it are not. Just how flawed this interpretive vision was in practice is something that no one could have calculated ahead of time because the spy plane program was still so untested.

This kind of thing was not supposed to happen. Unlike in 2016, the 2020 trial was public, and the program was presented as a highly regulated, data-driven, scientific test. This time, police promised, they would do the spy plane right. As I will show, however, even in its ostensibly more transparent 2020 iteration, the BPD actively downplayed crucial information about the spy plane's risks and potential harms. PSS, for their part, was incentivized by their for-profit relationship with police not to reveal to the BPD the limitations of the technology. In the end, as I will show, on at least one occasion, the spy plane mistakenly sent police to intercept Baltimore citizens who had no involvement in a crime. Just how common a problem this was, however, remains a mystery.

Total Transparency and Accountability

The spy plane first came to Baltimore in 2016 under total secrecy. As discussed in chapter 1, that first experiment quickly blew up in the BPD's face. The public skepticism from that time deeply shaped how police and PSS approached things in 2020. When it finally came time to announce the program was coming back to the city, then, PSS and the BPD tried to be more transparent. They wanted to make sure that those most directly affected by the city's gun violence, primarily Black residents of West and East Baltimore, were supportive.

Transparency and accountability are practices that the BPD is not exactly expert in. This showed in the ham-handed way they engaged the community. Police officials quickly turned to a tactic that, I would later learn, is part of a long-standing playbook for deploying new surveillance technology: don't give enough information about a new technology for the public to have an actual debate. The BPD would frame the program so narrowly that it was hard to ask informed questions.

Circumstances didn't help the rollout. It coincided with the onset of a global pandemic. With the program slated to launch in May 2020, the first swell of cases of COVID-19 were ravaging Baltimore just when the BPD was set to engage with the public, throughout late March of that year. In lieu of in-person sessions, which were banned at the time, Police Commissioner Michael Harrison chose to broadcast through Facebook and take questions from the public via online chat. Using PowerPoint slides, he presented a few still frames of grainy WAMI imagery, taking much of the description of how the technology worked almost word for word from PSS's promotional materials.

Few people attended these sessions, not surprisingly. For those who did, the presentations were confusing. Harrison stated that, on the one hand, the spy plane was part of an "evidenced-based and data-driven" strategy of policing, which would deploy the technology with scientific precision, guided by "data." Yet, in the same breath, he told citizens that "we don't know if the program will have any impact on the crime in our city."[1] At one of the public sessions, he said the spy plane was "an experiment that has never been done in an American city. . . . There is no expectation from us that it will work. This is a pilot program and we're guided by research."[2]

This was confusing to the online audience. How is this an evidence-based approach to policing if the technology has no evidence to back it up? One citizen commented in the Facebook chat stream, "Any proof that it actually works? Has [sic] any other places used it with success?" And if the point of the trial period was to *generate* the evidence that *could* tell the public if the technology worked, then just what kind of risky situation were they putting the public in by using this untested tool? As I would later learn by looking at the BPD's history of surveillance experimentation, this is familiar behavior. The BPD's idea of an experimental, scientific approach to policing has a pseudoscience feel, featuring attitudes toward truth that "do not play by the rules of science even though they mimic some of its outward features."[3]

Police officials obscured the facts about two main areas of risk when presenting the program to citizens: objectivity and racial bias.[4] Firstly, they downplayed the incredible amount of human labor that goes into linking the imagery to useful identifying evidence, such as CCTV footage. Harrison neglected to talk about the crucial communicative relationship between the plane's analysts and detectives, which would lead to an arrest. As the Pulaski case, among others, would later reveal, a lot can go wrong when this communication is not seamless. Responding to a question about the imagery quality, Harrison said,

> As you saw in the presentation, it's only one pixel per person, and so it's very grainy and so you won't be able to make out any physical or distinguishing characteristics of a vehicle or person. It will only track the movement. . . . We're tracking movement, we're not even necessarily tracking people.

This is confusing. How will the imagery be useful for identifying people if it can't identify people? If the imagery is so grainy, won't it be hard to use? What does tracking movement, but not people, mean? Later, Harrison noted that "we're not accessing . . . the data unless an analyst hired by the vendor [PSS] and working for the vendor gives it to us." What does "giving the data" mean? Giving how? Instead of answering these basic questions about how the public-private partnership would work, Harrison focused on PSS's often-repeated claims that the system is not invasive because it "can't see what you look like." He reassured listeners, "Because the plane can only download one pixel per person, you cannot make out any distinguishing characteristics of any person, or any distinguishing characteristics of any vehicle. . . . . It is only an investigative tool to be added to the toolbox for detectives to use."

Judging by comments in the Facebook chat, attendees were confused and asked sophisticated questions about why they should trust

PSS to interpret the imagery. "My concern is what happens to the data. Who is interpreting the data?" one person pointed out, adding, "That's problematic for me that the vendor is the lone interpreter of the data." "Why not offer the public a true simulation of how it works?" another person asked. Lurking in these short comments—there were many others like this—are high-level questions about the objectivity of the imagery and whether or not police or PSS analysts might make mistakes when interpreting the data.[5] What, exactly, does it look like for an investigation to progress based on evidence from the plane, especially if it is so grainy? Why should we trust the people interpreting the data?[6]

Secondly, BPD executives appear to have given almost no consideration to whether or not the program could be racially biased. When a citizen asked how the department would "ensure that racism and racial bias is not a part of this program," Harrison appeared flummoxed. He pointed to the department's willingness to be subjected to an independent auditing team, composed of highly respected entities like the NYU Policing Project, to make sure the technology would be used ethically. He responded,

> Did the analysts do the download the appropriate way? Was the case file put together the appropriate way? Was our investigation by the book? . . . So if [auditors] catch us doing that [in an unethical way], we're open to them telling the community whatever it is they find. That's how confident I am we won't be using it in any nefarious way—that we won't be manipulating it to satisfy our own purposes.

Harrison's construction of racial bias as the result of "nefarious" behavior is telling. It reflects what scholars call an individualistic, as opposed to structural, understanding of racism.[7] The only way the program could unevenly, negatively impact Black Baltimoreans, he seemed to be saying, was through the actions of individual "bad apple" officers, and this would be rooted out by auditing. He said nothing about whether and how the department would prevent the technology's built-in risks (about which little was known) from being shouldered most heavily by those living in Black neighborhoods.

Harrison also presented the program as race-neutral by talking about the imagery's breadth. Not only could the plane's cameras not see skin color, but the plane would be able to look everywhere and at everyone. In fact, as McNutt also liked to emphasize, it would actually capture imagery of White neighborhoods *more*. On the surface, this was plausible. PSS would fly two planes, one with its orbit centered over the Westside and the other over the Eastside. Overlapping orbits in the center of the city, the total system would be photographing some of the wealthiest and Whitest neighborhoods in Baltimore's central and northern districts *twice* per second. In this sense, it was actually White people who would be made more visually available to the state's gaze. As I will show, this is actually a misleading explanation, but the public presentation didn't go into enough depth for community members to understand.

In the end, even though attempts at public transparency were derailed by the COVID-19 pandemic, I suspect that the BPD had already fully committed to the program by the time of these sessions. They were a performance. Indeed, before these public hearings commenced, PSS was already putting up drywall and installing monitors in the operations center. The sessions were ultimately held to defuse public criticism under the guise of consultation. The few citizens that did turn up asked great questions, few of which were answered, and much confusion remained by the time of the first spy plane flight.

Racial Bias

Contrary to Harrison's assurances, in my view, the spy plane program was racially biased. To see this, however, you have to know what was going on behind the scenes. Most of the evidence of bias is absent in the public record.

Orbits and Image Quality

For a host of technical reasons, spy plane imagery quality is not uniform throughout. It becomes fuzzy and difficult to decipher at the edges. As the plane circles an area, the locations at the center of the plane's orbit, farthest from the plane itself, are photographed the clearest. McNutt had to decide, then, exactly where to center the orbits of each plane. Where should the best imagery be available? He chose East and West Baltimore because these are known violent crime "hot spots." Where the imagery is the worst, then, would run along "cold spots"—a strip of wealthy, majority-White neighborhoods in the center of the city with less violent crime. Double the blurry imagery does not allow analysts to see White neighborhoods better. The assertion that the planes watch White people more seems to have been a political clapback to "woke" critics, not a description of reality.

In my view, the question around orbits and image quality is a distraction. It plays into what I have been calling the "doomer" critique, where we assess the harms of policing technology in terms of its threat to "watch us all." Instead, we need to center the problem of experimentation. From the outset, we should assume the spy plane is like many other policing gadgets: a glitchy, pseudoscientific tool with unforeseeable risks. Where were the mysterious risks of conducting aerial surveillance with this untested technology felt the deepest? Whose lives were used to test out the plane?

Mapping the Eye in the Sky

Based on internal data from PSS, I mapped the incident locations of all 180 investigations conducted by spy plane analysts in 2020. Each

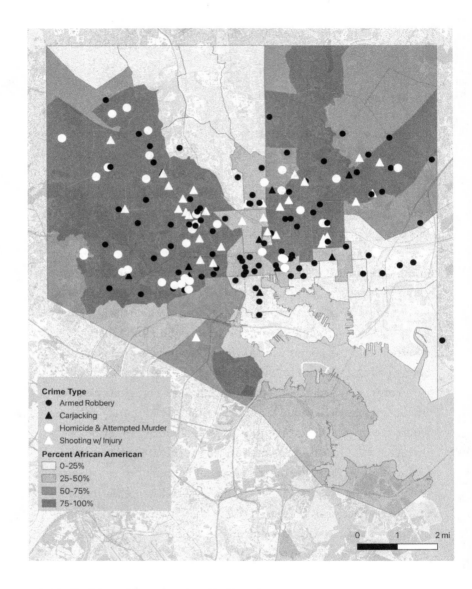

MAP 1. Racial map of spy plane investigations, 2020.

mark in map 1 represents the starting point from which analysts began following people and cars in their investigations. Laying these data over a racial map of the city, which illustrates the clear divide between majority Black and White neighborhoods, yields a pattern. Most investigations were in predominantly Black neighborhoods. Breaking these investigations down by crime type, you can see that, when analysts did conduct investigations in White neighborhoods, they were (save for four homicides and four shootings) typically related to property crime (robberies and carjackings) in the central business district.[8]

Instead of seeing these marks as evidence of racial bias because they show where the state concentrated its "creepy" panoptic gaze, however, I suggest we look at them as "doses" of a pseudoscientific intervention. As I will argue, like many pseudoscientific techniques used in law enforcement, the spy plane was deployed in such a way that the tool's harms could not be easily assessed.[9] When analysts tracked, they exposed to these harms not just those who were involved in the crime, but anyone else who just so happened to be in the area. In this sense, only certain neighborhoods received an outsized dose of an untested drug. This leads me to ask, then, what kinds of risks were residents of Black neighborhoods exposed to? Were PSS and the BPD prepared for those risks? Did any concrete harms emerge as a result of those risks? If so, were they recorded and reported, and was anyone held to account?

Making Mistakes on Black Lives

The moment I learned about the Pulaski Effect, and saw for myself just how easy it is to convince oneself that something completely normal is "suspicious," I was waiting for the other shoe to drop. It seemed like only a matter of time before one of these misidentifications would put an innocent person in harm's way.

I should have recognized the problem sooner. In one of the first interviews I ever conducted with PSS personnel, one of the company's original analysts related a story about misidentification that should have disturbed me more than it did. Back in the early 2000s, PSS was given a trial contract with U.S. Border Patrol to surveil the US-Mexico border. One time, they caught sight of a long line of people making their way across a scrubby patch of desert in a single file line, which they assumed to be a train of drug smugglers. "We saw that, and that checked every box for, okay, this is a drug mule . . . carrying potentially a large amount of drugs on their back moving through very difficult terrain," the analyst recalls. Tracking them in real time, they sent agents to intercept the train as it came to a back road. When the agents arrived, however, they were surprised to find nothing but a herd of cattle grazing, single file, through a patch of wilderness. PSS had sent Border Patrol after a herd of cows. "It was crazy," the analyst reflects, "because we were so sure that like this is the one. This is gonna be our perfect example of how we can help here." The Pulaski Effect strikes. The emotions of being able to close a big case and impress law enforcement got in the way of objective analysis, making the blurry imagery even easier to misinterpret. The analyst recalls that the mistake was "deflating at first," but he and others took it as a learning opportunity. "You can't be too sure," he said. "You can't assume identity." Over time, the story became a kind of humorous cautionary tale that was told to trainees to encourage them to never make assumptions. Clearly, then, PSS had long understood that subjective perception is a big factor in the reliability of spy plane analysis.

This story should have given me pause, but, like everyone else who had gotten seduced by the idea of an all-seeing eye, I must have harbored a deep faith that the imagery was fundamentally objective. Surely this was an outlier. In a big operations center like the one they had in Baltimore, with so many people looking at the imagery, double-checking each other's tracks, with multiple layers of oversight, surely the objectivity of the camera lenses would overcome flawed human eyes. The truth would always out itself. Right?

The other shoe did finally drop. Though I was not present in the operations center to see the case firsthand, through internal documents and interviews with those involved, I was able to piece together a moment when PSS came within a hair's breadth of triggering the false arrest of a Baltimore citizen.

The McHenry Street Case

In late fall of 2020, just a few weeks before the program was shuttered, a shooting occurred on McHenry Street in a neighborhood called Sandtown—the heart of West Baltimore—around ten o'clock in the morning. Detective Karim, an experienced investigator who had been a frequent user of the spy plane, arrived on the scene. He told me he was immediately given multiple witness statements identifying the shooter by name—a rare development in a murder case in this part of town. Karim immediately ran the suspect's name and found out he was on federal probation, which meant a large file of identifying information was already available. The suspect was known to drive a distinctly recognizable vehicle: a dark blue Cutlass. Karim ran the Cutlass's plate through the city's license plate reader (LPR) database and got a hit. About nine minutes after the time of the shooting, the suspect's car passed by an LPR camera just a few blocks north of a public housing complex on the Eastside. The timing was right to suggest that the suspect had been driving away from the crime scene that day right after the shooting.

Right away, then, Karim had a good case going: he had a lead on who did it, on what the suspect drove, and that the Cutlass was heading away from the shooting scene shortly after it happened. Still, Karim had no concrete evidence to link the Cutlass to the crime scene

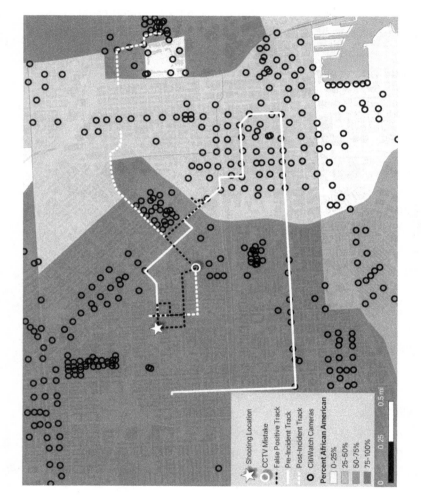

MAP 2. Tracking in the McHenry Street investigation.

and to confirm the suspect as the actual driver of the Cutlass. He turned to the spy plane for answers. Maybe PSS could find a vehicle of that description around the crime scene. If they got lucky, they could even track the car to a place where the driver got out and pull ground camera footage of the suspect driving the car. That would be more than enough for an arrest. Karim sent in a no-frills request with the address and time of the shooting.

The case fell to Team 1. They began their investigation in typical fashion, by tracking all the cars on the street where the shooting occurred, which amounted to something like six vehicles.[10] Right after the time of the shot, they saw people flee on foot and in vehicles—a clear sign of a gun crime's aftermath—and tracked them all.

Analysts immediately ran into what would become the biggest problem for this case: the crime scene was located in a CCTV desert. As illustrated in map 2, there were no CitiWatch cameras within blocks of the location in any direction. It's important to pause on this point because it's not just a coincidence. The fact that there were a lot of cameras *around* this shooting, but none right next to the crime scene, is part of a larger historical pattern. As I explore in the next chapter, Sandtown lies just to the west of one of the major racial "borderland" areas between Black and White Baltimore. These borderlands are thick with CCTV cameras because of the critical part they have played in shoring up long-standing lines of racial segregation. There were cameras on almost every corner just five or six blocks in every direction from the McHenry Street crime scene, but nothing close by.[11] This was so despite the fact that this neighborhood was a clear gun violence hot spot. In 2020 alone, Sandtown saw fifteen shootings and nine homicides.

Finding themselves tracking in a camera desert, analysts had a tricky task: any car that left the crime scene would need to be tracked for many blocks until it just so happened to pass a CCTV camera somewhere in the surrounding area. Then, they could check to see if it was the dark blue Cutlass. They continually came up empty. No departing vehicles matched the suspect's car.

After about a day of dead ends, Karim finally made an important discovery: he found security camera footage from a corner store near the crime scene. The camera didn't capture the actual shooting, but did show a person running on foot, gun in hand, in the direction of the shooting. PSS analysts realized they had been wasting critical hours tracking cars on the street of the crime scene, when the suspect probably hadn't even parked there and had approached the scene on foot. Karim described these first few critical hours of the investigation, in which analysts went on multiple wild goose chases, as a "time suck."

Having been directed to drop the task of tracking vehicles at the crime scene and instead try to track a person on foot, analysts quickly found their mark. Following a tiny smudge/dot of a figure eastbound on McHenry, they tracked the suspect running at high speed rounding a corner onto a street just a block over from the shooting. The person then disappeared in the dark shadows cast over the sidewalk of a very narrow street. Just a few seconds later, though, a vehicle parked midway up the street left the scene. According to Karim, with no CCTV in the area to help, analysts couldn't tell him if the person they tracked on foot got into the vehicle, but the coincidence of the movement between the person and the car was enough that analysts operated under this assumption. They had to make many assumptions. The question became whether this car matched the one associated with their primary suspect: a dark blue Cutlass.

Sending the Police to the Wrong Address

As with every time-sensitive investigation, to speed things up, analysts broke the tracking of the target car into two: a pre-incident track, in which the vehicle was followed backward in time, and a post-incident track, which followed the target forward in time to see where it ended up. The forward track (the dashed white line in map 2) progressed flawlessly. Analysts tracked the car eastward through the racial borderlands beyond Sandtown, into the Eastside where it matched the timing of the LPR hit. They could now confirm that, indeed, the car that departed the scene (which presumably was being driven by the person who fled) was the same dark blue Cutlass captured in the LPR. Because it went through the borderlands, passing many other ground cameras, it was easy to verify the look of the car and the continuity of the tracking time and again. They even lost the car in a particularly wealthy area right in the center of the city (note the gap in the dashed white line in map 2), but picked it right back up again as soon as it passed by another borderland camera on the Eastside.

Despite the analysts creating a solid post-incident track, an important piece of information was still lacking. What was the suspect doing before the shooting? Sure, they could prove his car left the area *after* the shooting, but the suspect could easily deny being at the crime scene. And what if the person who fled on foot didn't actually get into the vehicle, as they had assumed? Unless analysts could prove the suspect was *also driving* the Cutlass *before* the shooting, the case might fall apart. Determining that crucial detail would be the job of the pre-incident track.

The task fell to Sarah. Her work is represented by the solid white line and dashed black line in map 2. Right away, she ran into problems. The street where the car was parked was covered in shadow and very narrow. Cars on the street were almost unnoticeable in the footage. What was more, because the buildings stuck out above the surface of the ground, every time the spy plane circled in its continual orbit, these buildings appeared to lean over the street for a few minutes, blocking her view of what was going on. As the plane came around, the view of the street opened up again for a few minutes, like a maddening aerial peep show, at which point Sarah had to refind her target car each time. Was it still there, or had the car pulled away in those few seconds when she was waiting for the street view to open up again? Finally, after a few rotations, she was lucky enough to see the car move away. (Because she was tracking in reverse time, the car was actually arriving and parking just prior to the shooting.)

Sarah continued to track. She knew this was the Cutlass because she picked up the target from the post-incident track, which was rock solid. But as the car moved farther back in time, going through many tricky spots with shadowy streets and narrow roads, Sarah became desperate for another ground camera, just to make sure she hadn't accidentally "jumped cars"—that is, accidentally clicked on the wrong car when it was temporarily obscured.

At this point, I am not sure precisely how Sarah's tracking proceeded because this information is not preserved in internal documents. Based on interviews with her and Detective Karim, the following is a good approximation of what happened and is represented by the dashed black line in map 2.

Block after block, Sarah tracked the car as it made a strange winding route a few blocks south of the eventual crime scene, still well beyond the borderlands and thus no ground cameras to help. As Sarah recalls, eventually the car came to a complicated V intersection. It paused, about to make a left turn against busy traffic. There were dozens of cars in the area, which meant Sarah needed all her concentration to keep eyes on her car and not accidentally jump to the wrong one. And then, the worst happened.

The imagery went black.

Because spy plane footage is actually a composite of twelve individual camera images that are digitally stitched together, sometimes the "stitching" is less than perfect. It's glitchy. The problem especially plagued the second of the two planes. That system had to be built in a rush because of supply chain delays, caused by the COVID-19 pandemic, and was never properly tested before deployment. It was continually producing glitches like this.

The glitches were unforeseen. Then again, this was all part of PSS's approach of "rapid prototyping." We build the plane while it's flying. Glitches are a feature, not a bug, of the program's deployment philosophy.

The bad stitching created big triangular-shaped blocks of black nothingness. These black wedges spun in random patterns as the imagery refreshed each second. In this instance, it just so happened that one of these stitching gaps fell right over the top of Sarah's target vehicle at precisely the wrong moment. "[The imagery] just kept going black, black, black [with each click] but, like, kept moving where the car was supposed to be," Sarah recalls in frustration.

Using all her skills as a tracker, she did what analysts always do in these tricky situations: count cars. As described in chapter 2, analysts are taught that just because you lose sight of your vehicle doesn't mean it is lost. You simply determine the pattern of traffic around your car prior to the problem area, and then see if you can find your car in that same pattern once it reemerges. Sarah did just that and was able to find a car that she thought might be a match, but something told her it might be wrong. She says,

> I think it was just kind of this gut feeling. Like it's just little things didn't look the same on iView anymore. . . . When you've been tracking with cars for a while, like the same car, you really get to know it in all the shadows, and on shady streets and on the bright streets. I kind of was just feeling like I'm not sure.

With her confidence shaken, Sarah was desperate for corroborating footage that could tell her if she was still on the Cutlass.

Then, a stroke of luck: the car finally went a few blocks south and east, just far enough to enter the beginning of the borderlands on the edge of West Baltimore. It went past a ground camera that lies just on the edge of West Baltimore (a large white circle in map 2). Sarah, wanting to save time and speed things up, asked for help. She turned to her partner to check the ground camera for her, to see if it was the Cutlass. She recalls,

> I said, "Hey, can you look at this ground camera, the car passes by it, and can you just tell me when you see the Cutlass?" Yeah, so [my partner] looked through it and was like . . . "No," she's like, "It's not there." So, I took that [to mean] the camera rotated.

Sarah had misunderstood. She thought her partner was telling her the camera "missed" the Cutlass because it was turned away, when in fact the message was that no such car passed by. This is an easy mistake to make. Most of Baltimore's CitiWatch cameras "pan and scan," so it is not uncommon for a car to pass a ground camera as it is looking away. That's why a track needs multiple ground camera passes before it can be deemed solid. But in this case, the "no" actually meant it was "not possible" the Cutlass was in the area.

Having misunderstood her partner, Sarah continued tracking the car. She watched as it finally pulled into an apartment complex just on the edge of West Baltimore.[12] This looked like a potential start location for the suspect's journey and a perfect place to see if the suspect was actually in the vehicle prior to the shooting.

Sarah pulled Zach over to help her check the tracking. She recalled, "I was like, 'Please . . . I really, really, think I jumped.' And he's like, 'You didn't jump.' . . . We watched it over and over." Zach and a few others on Sarah's team looked over her track several times, and each person thought she had successfully followed the same car through the area with bad stitching. She said, "We were sitting there and [Zach] is like, 'Don't doubt yourself.'"

Everyone was convinced Sarah had the right car partly because its movement looked "suspicious." According to Sarah (there is no internal documentation of this), the mistaken car went around in circles, back and forth, along different blocks within a small area just south of the incident location (simulated by the dashed black line in map 2). Analysts had seen this sort of driving behavior before when suspects were looking for their victim or scoping an incident location. "The car made sense," Sarah recalls. To the analysts' eyes, the fact that this car was in a known "high-crime neighborhood" and wasn't just driving from point A to point B made it suspicious. Similar to how patrolling officers have more discretionary power to stop and frisk people for walking in a neighborhood that the officers deem "high crime," analysts' judgment was colored by the area of the city in which they were tracking.[13] Even though Sarah was not feeling certain, then, these kinds of racially coded biases about West Baltimore neighborhoods partly overrode that feeling. Going against her gut, Sarah recalls, "I kind of gave the address where this car went to Zach, and I was, like, 'Look there's potential—I'm not one hundred percent sure, but there's a potential possibility that the guy went here.' And he right away *messages* Karim." Sarah says this with a "can you believe it?" sort of tone.

Lacking other good leads, Karim drove to the apartment complex to see if he could find camera footage of the suspect in the Cutlass, or any other information from residents. "We were harassing this apartment complex for forty-five minutes looking at footage [of security cameras]," he recalls, "and asked [the superintendent] about who lives there, and where, and that kind of thing. And [*pause*] it's the wrong person, obviously." Karim even called PSS and had analysts lead him as he drove around the area looking for private security cameras that, in iView, made it appear that the Cutlass had passed. No corroborating footage could be found. In addition to the apartment complex, Karim talked to owners of liquor stores and corner stores and couldn't find footage of the Cutlass anywhere. "So, was it a waste of two days of my time?" Karim reflects. "Yeah. It was awful. Was I intruding on this person's privacy? Of course. Yeah. But it's not—it happens." It is unclear exactly what he means by "harassing," but the fact that Karim spent nearly an hour at one of these locations asking questions and "intruding on" the "privacy" of at least one person suggests that this may have been an unpleasant encounter with police.[14]

Discovering the False Positive

While Karim was out knocking on the wrong doors, Sarah continued to have a sinking feeling that there was something wrong with her track. She decided to go back to the one ground camera the car passed, which she thought her partner had told her was inconclusive. She pulled up the footage and was shocked to see that the camera doesn't pan and scan. It is a fixed camera. "The camera doesn't rotate," she recalls. "It's just [showing] a different car." Her partner was saying a Cutlass could never have passed by the camera at all!

Panic!

"I was like, oh man, freaking out. You know, I was like, 'Oh, we got the wrong address, I got the wrong address,'" she said. She immediately got in touch with Karim and told him, "Oh, looks like perhaps he doesn't go to that address." Karim stopped his search before he could "harass" (to use his word) any more citizens.

Karim told me he was not terribly surprised by this mistake. "That's what happens when you look [in] real time," he says. "Mistakes get made. It's not the end of the world." He then tells me about many other examples like this. He says, "That's gonna be every time with every technology. We get the wrong, sometimes we literally, you know, a typo gets us the wrong person's cell phone.... I have personally subpoenaed the wrong person's cell records." Karim's mind immediately goes to cell phones when thinking about the kind of false positive in the McHenry Street case. As I discuss in the next chapter, much like cell phone tracking techniques, the spy plane's false positives expose lots of people to mistakes because it can capture such a wide range of individuals surrounding the target of interest. For Karim, then, this event was unremarkable. He sees it as part of the process of conducting time-sensitive investigations with experimental technology. It was very inconvenient for him, though, and bogged the investigation down with useless information.

Sarah had similar feelings about the mistake. She reflects,

> Where we tracked was like very, very tricky. I had so many people helping me with it, and, like, it was obviously disappointing and frustrating to give the wrong information, but at the same time, like, you learn from that, you know, just kind of be confident in what you are [confident about] and not what you're not so confident about.

As with Karim, for Sarah this incident was less than ideal, but ultimately struck her as a routine mistake that can happen when you're in the business of tracking murderers. The more important thing is to maintain your confidence. By her own accounting, then, Sarah sees spy plane analysis as more of an art than a science—it's about gut instincts, confidence, and feel. So much for being a mechanical witness...

A Second False Positive Is Found

At this point, Sarah figured she could go back, find her mistake, and still try to provide some pre-incident movement for Karim, even if the case was getting cold by now. She enlisted the help of Rania, who had become PSS's "fixer." Just like when she was a newbie on the Pulaski case (and you may recall her heroics on the Carroll Avenue case from chapter 2), she could always figure out a botched track.

Rania went back to the tricky section where the image stitching caused Sarah to jump at the V intersection and somehow managed to see the mistake. She fixed the track onto the correct car. When Sarah and Rania tracked the car to a nearby ground camera, they were shocked by what they saw.

This car was not the Cutlass, either!

Not only had Sarah jumped cars at the area with bad stitching, sending Karim to harass random citizens, but it seems she had jumped cars even earlier. The entire track (the dashed black line in map 2) is wrong. Even if Sarah had not jumped because of the bad stitching, then, she still would have ended up triggering a false positive. Even her "gut" hadn't told her anything about this earlier mistake. Sarah recalls,

> I was like . . . I was so flabbergasted because I was like checking my tracks from point A to where I lost him . . . a thousand and five hundred times, like, over and over and over, and I was like asking supervisors to look at it—everybody—and the tracks were perfect. It just wasn't the right car.

The more certain you are, the less right you are. The Pulaski Effect strikes again.

With sharp-eyed Rania by her side, Sarah reviewed the analysis all the way back to the incident location, where she had originally picked up the backward track. It was there that Rania discovered the original mistake. While the car was parked on the street just prior to the shooting, with the buildings doing their infuriating peep show and the shadows constantly obscuring things, Sarah clicked on a car parked right next to the Cutlass and didn't notice. One mistaken click at the beginning of the track triggered a cascade of mistakes that put innocent Baltimoreans in the crosshairs of a detective.

The Mistakes Become a Success

Sarah and Rania decided to track the whole thing again and finally get it right, even though Karim had surely lost all patience with them by now. They began near the crime scene again, this time on the correct car, and started to build a long track (the solid white line in map 2). "We really, really tracked it all—as far as we could 'til a camera [*pause*]. We found it in a camera. That was it. That was the vehicle," said Sarah. Finally, they really had the Cutlass (really this time) and were able to confirm it in a dozen cameras, but only because they painstakingly tracked the car way back in time, beyond the "camera desert" of Sandtown and into the racial borderlands.

Finally having a positive confirmation that the track was right, Sarah and Rania kept going. They tracked the Cutlass as it doubled back west of the crime scene. And then . . . jackpot. It visited a gas station, more than an hour prior to the shooting. Karim pulled the footage from the station's high definition security cameras, which showed a person getting out of the Cutlass. The figure was wearing similar clothing and had a similar build to the person in the footage fleeing the crime scene. For Karim, this visual similarity was slam dunk evidence.[15]

Taking this information into the interrogation with the suspect proved crucial. The suspect denied being present at the shooting, denied even being in the area, but Karim found a way to use the gas station evidence as a tricky work-around. He recalls,

> He doesn't want to say he was anywhere near the crime scene, obviously, but he would, when I asked him about, you know, "Maybe did you go get gas, or did you have a problem with your car?" And he was, "Oh, yeah"—I pull up photos [from the gas station footage]—he's like, "Yeah, I was putting air in my tires." That's clearly him. He knows it's him. He saw himself in the camera footage. . . . Why would you care at all that we have camera footage at a gas station an hour and a half before a crime, you know, a mile away from the crime scene? Normally . . . you can have a person do that all day and it doesn't help you [close the case] at all, right? In this case, because we can directly link that person there [in the gas station footage] to the

crime scene, that was more than enough. So, it gave us—he refused to confess to [the shooting], but he gave enough that it didn't matter.

Karim arrested the suspect for murder. In the end, Sarah and Rania's fixed track closed the case, not so much because it provided a "ground truth" of the crime scene, but because it provided a piece of circumstantial evidence that could trip up the suspect during interrogation and thus indirectly lead to a partial admission of guilt. Even though the spy plane didn't "work" as advertised, even though mistakes were made that invaded people's privacy, and even though it wasted days of his time and bogged down the case, from Karim's perspective, it was still useful. "It's not a be-all end-all," Karim reflects. "Everyone thinks it's better than it is. . . . But it does help link things."

What's the Harm of False Positives?

Based on Commissioner Harrison's claims about the professionalism and unprecedented rigor and transparency designed into the spy plane experiment, one would expect that the kinds of limitations exhibited in the McHenry Street case—problems like narrow streets, shadowy streets, communication errors, "stitching" glitches, and so on—would have been caught by auditors and reported up the chain of command. Or, at the very least, some sort of record of false positives would have been kept in order to quantify the risk. The spy plane can clearly result in false positives, especially in areas with tall buildings and no CCTV, and this seems to have happened more than just one time. Okay, well, how many times? Nobody knows. No one inside BPD management, city leadership, or even the Policing Project (the ethics and privacy auditors) were made aware of this incident or other examples like it. Even the company's internal documents of this investigation make no mention of the two false positives and Karim's unwarranted questioning of citizens.[16] If the suspect's defense attorney looked at the briefing document disclosed to him by a prosecutor, there would be no record of these mistakes. If a journalist requested the briefing through the FOIA process, none of this information would be seen. All that was saved for the record is the final product, all shiny and perfect, with clean tracks. It was only through interviews with the people who were there that I was able to reconstruct the events.

One reason false positives didn't cause more of a stir and leave more of a paper trail might be because they were seen as normal. To be clear, I don't think analysts and detectives were trying to cover up their mistakes. There was nothing "nefarious" going on, to use the commissioner's term. I think they just didn't see them as egregious errors. Mistakes happen when you're engaging in experimentation, investigators thought, and the mistakes were caught in time. Moreover, at least with the McHenry footage, the case ended successfully by providing evidence that helped put a suspect in jail. Surely that outweighs any costs that might have been incurred by Karim's harassment of citizens, they might say.

I have a different view. One reason these mistakes didn't leave more of a mark on the program is because of the financial incentive to make sure the technology is perceived as flawless. As Cathy O'Neil has observed, this is all-too common within high-tech organizations. Designers often fail to gather data about false positives (and negatives), and their processes are then ill-equipped to factor them in, self-correct, and reduce bias.[17] This is related to the profit motive structuring the organization's goals. For-profit start-ups often neglect, or choose not to measure, things that will threaten their bottom line.[18] If part of start-up culture is to always focus on the positive, "fake it 'til you make it," then speaking out about or even keeping track of the flaws and glitches can become an almost intentional blind spot inside the company's operations.[19] The tendency to have hidden limitations that no one wants to talk about may be a feature, not a bug, of for-profit technological experimentation.

The Harm of Unforeseen Risks

How are we to judge the kind of harm, if that's what we want to call it, that came from the Pulaski Effect? The least critical take would sound something like this. The spy plane program was an experiment, and, as with any experiment, it revealed some unforeseen risks. These could only be discovered by actually deploying the technology in a real-world situation. Accidents happen, even with the safest technologies. A few false positives are outweighed by the fact that the plane can occasionally break open a case, including in the McHenry case. If the spy plane is "just another tool in the toolbox," as Commissioner Harrison often called it, then it is a mostly safe tool, some might claim. As the saying goes, don't let perfection be the enemy of the good.

I disagree with that take. While a few false positives might not be so bad if a start-up is tasked with, say, targeted advertising on the web, it is far too big a risk when tasked with helping put someone in prison. PSS needed to be held to a higher standard than other tech companies. To put it bluntly, the "drug smuggling cows" debacle from early in PSS's development, as well as the Pulaski case itself, which happened in the first month of the trial, should have hit the company as cautionary signs of a serious risk around which a rigorous training and monitoring program was built. BPD executives didn't even know that false positives were happening. It was the same for attorneys. In my interviews with both the prosecutors and defense attorneys who worked this case (discussed more in chapter 7), no one was made aware that these mistakes happened. The only plan PSS had for false positives was to "always get a second opinion." As Sarah demonstrated, though, she got second, third, fourth, and more opinions, and that still didn't prevent the mistake from happening, not once but twice. Such a lack of rigorous safeguards against confirmation bias is a hallmark of pseudoscience.[20] At the very least, like other professionals who conduct experiments, such as medical experts, there should have been a mechanism to check for and gather data on false positives so the public could be made aware of the risks, should the program be extended. As I discuss in chapter 8, there are good examples for how to do this, but it requires police executives and community stakeholders to be able to distinguish between legitimate and illegitimate forms of experimentation, and to force tech companies to operate at a higher level.

The Harm of Unequal Harm

The McHenry Street case reveals an even bigger problem, though. It is not just that false positives were made by the spy plane, it's that there was no recognition within the company or the BPD of how mistakes like these play into Baltimore's racialized surveillance history. The mistakes resulted in sending a detective into an apartment complex in a Black neighborhood on false pretenses. Thank goodness Karim's attempt to locate the mistakenly tracked vehicle didn't end in a more violent form of harassment. I am certain, though, that such an event was only a matter of time. How many more cases would need to be worked before police actually violently engaged an innocent Baltimore citizen because of a false positive from the spy plane? Whatever the answer, you can be almost certain that such an incident would happen in Black Baltimore. The program gave an outsized dose of its experimental surveillance to race-class subjugated neighborhoods, so it is residents of those communities who bore the risks the most. It would only take one, just one, really bad incident to traumatize those communities yet again, dredging up the long history of harmful experimentation that generated much of the distrust toward the BPD in the first place. In this way, the spy plane could actually create more distrust toward police. Yes, "it could have been worse," but is that a risk worth taking in a neighborhood that has already experienced so many harms from hyper-policing? The risks of experimentation do not have the same political consequences for all Baltimoreans.

There was no public debate about this kind of structural, racial inequality before the planes were launched, even when citizens asked (via Facebook chat). Officials were too busy trying to make the spy plane sound less creepy by reassuring them that it "can't see what you look like." Who cares if it can't see what you look like? In fact, it turns out, that might end up getting you falsely arrested.

There is still one thing bothering me about the McHenry Street case. It turns out that the Sandtown neighborhood is a nightmare for trackers—a "tricky tracking area," to use Sarah's phrase. Yet, this is precisely the neighborhood where the spy plane was advertised as a magic bullet for crime. How can this be? How can the spy plane be both the prescribed antidote to West Baltimore crime *and* so unreliable when used in precisely this area? As I explore in the next chapter, this is how it has always been in Baltimore, unfortunately. It is far from the first time a supposedly game-changing crime prevention tool has been discovered to be glitchy or dangerous when Black neighborhoods were used as a petri dish. The spy plane sits in a long history of collateral damage from failed experiments. That's the lesson I take away from the McHenry Street case. Putting a wide-angle lens in the sky is not enough to overcome the decades of race-class subjugation baked into Baltimore's geography; in fact, it might just make it worse.

4 Experimenting on the Black Butterfly

People often grimly joke that Baltimore is two cities: Balti*more* and Balti*less*. Lawrence Brown, a public health scholar who spent years studying Baltimore's social geography, has dubbed these two cities the "White L" and the "Black Butterfly."[1] So powerful have Brown's concepts become, they have seeped into everyday language in Baltimore. You might hear folks in the city use them in casual conversation. White citizens cluster tightly in a central L-shaped corridor that runs north/south along Interstate 83 and then fans out from the central business district eastward along the stunning waterfront. Black citizens live predominantly in two regions that resemble a beautiful butterfly with outstretched wings: the Westside and Eastside. The L and the Butterfly, as Brown has documented, are remarkably good proxies for understanding the spatial distribution of just about everything in the city. Property values, private lending patterns, public transportation infrastructure, abandoned buildings, structure fires, toxic lead exposure, access to safe cycling lanes, even the locations of shared car services, all tend to follow this pattern of racial and class segregation.[2]

Like a lot of highly segregated cities, there are distinct lines of racial and class separation that can feel uncannily sharp. I think of these areas as racial "borderlands"—places where the lines of segregation are constantly being negotiated and thus where racial tensions often run the hottest. These borderlands, as I will show, are where surveillance experiments are typically conducted in Baltimore.

Crossing Greenmount Avenue in the north, for example, which divides the mainly White and wealthy area around Johns Hopkins University from several low-income Black neighborhoods to the east, means a difference in $30,000 per household in lending from banks separated by only thirty feet of pavement.[3] Eutaw Place, to take another example, runs diagonally between Westside neighborhoods like Sandtown, which sees dozens of murders a year, and Midtown, which sees one or two. On the Eastside, Eastern Avenue and Patterson Park form a wall between predominantly White neighborhoods with Baltimore's most expensive restaurants, such as Fells Point and Butcher's Hill, and predominantly Black and Latinx neighborhoods, like McElderry Park, which outsiders sometimes flippantly call "food deserts."

The two Baltimores and the specific locations of these borderlands emerged in large part through concrete legal and economic mechanisms that favored wealthier White citizens (less so for working class Whites) and disadvantaged Black citizens (*including* those who were middle class and upwardly mobile).[4] These mechanisms have now been well documented by historians of Baltimore and can be seen in nearly all of America's major cities: redlining, blockbusting, forced displacement, racial covenants, racial zoning, biased public housing policy, tax exemptions, subprime lending, undervaluation, and outright White vigilante terror upon (especially upwardly mobile) Black families have all caused cities like Baltimore to diverge racially and spatially.[5] To say that a city like Baltimore is segregated is to say that a dominant group—usually White property and business owners—has used legal and illegal mechanisms to leave subordinated groups—usually working class Black and Latinx citizens—with few options but to live in economically depressed neighborhoods. Whenever these racial groups have tried to migrate within the city (and they have *many* times), they have been pushed back, often violently.[6] That's why I prefer to use the phrase "subjugated communities," rather than the more passive and indirect language of "disadvantaged," "marginalized," or "low-income."[7]

Segregation is one of the primary things that fuels violence, particularly gun violence. Take a moment to look at a map of all the homicides by shooting in Baltimore in 2020 (map 3). The locations of these incidents have been laid over a racial map of the city, showing the Black Butterfly and the White L.

Murders overwhelmingly cluster in the Butterfly. The differential exposure to violence cannot be overstated. In 2020 alone, for example, a person living in Carrollton Ridge, a roughly 2,000- by 1,500-foot section of West Baltimore where the median income is around $19,000 a year, was surrounded by fifteen separate shootings that resulted in death, four of which involved two or more victims. Meanwhile, in that same year, the residents of Federal Hill, a similar-size neighborhood sandwiched between Baltimore's famed major league baseball park and picturesque harbor, with a median income over $120,000 a year, saw three shooting deaths. It has been like this for generations.

Segregation doesn't cause violence in a simplistic way—that is, when you get people of one race together, or when poverty is concentrated in one neighborhood, people start shooting each other. Rather, segregation is what sociologists call a "forming cause," like how a glass pitcher causes water to take the shape of a pitcher.[8] The glass doesn't have this effect by pushing on each individual molecule of water in the same way. Each molecule alters its position ever-so-slightly in a complex network of connections that ultimately leads to the water being shaped like a pitcher. Segregation is linked to violence through a similarly multicausal network—from the impact of

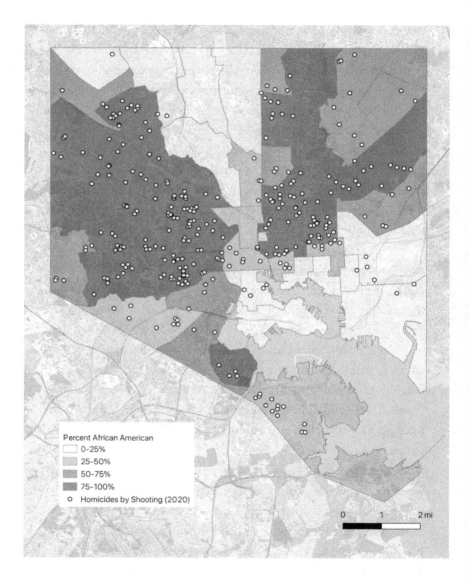

MAP 3. Racial map of homicides by shooting in Baltimore City, 2020. (Data from Open Baltimore.)

defunded schools, to the concentration of toxic lead exposure, to the concentration of abandoned buildings, to the concentration of violence itself. Segregation creates a kind of vicious, path-dependent cyclicity to violence, where the presence of heightened violence depresses the neighborhood even further, making conditions even more fertile for continued violence.[9] Violence that is shaped by segregation, then, is incredibly hard to figure out, let alone "fix," because it is not driven by just one (or even several) thing(s).

Segregation also shapes culture, particularly for wealthy White people. Replicated over generations, neighborhood racial isolation gives rise to what Black studies scholar George Lipsitz has called the "White Spatial Imaginary." "Decades and centuries of segregation," he writes, "have taught well-off communities to hoard amenities and resources, to exclude allegedly undesirable populations, and to seek to maximize their own property values in competition with other communities."[10] When cities stay so divided for so long, a certain cultural paranoia sets in among those who are benefiting, which leads them to want to preserve the purity of their neighborhood against perceived threats to prosperity—especially violence. In the White Spatial Imaginary, crime, even the threat that it *could* happen, must be kept out at all costs.

As Lipsitz notes, it is important to distinguish the White Spatial Imaginary from individual White people. The imaginary does not sit inside any one person's head (then we would just say "imagination"); it is more like a set of shared cultural assumptions about the best way to secure a properly ordered, prosperous neighborhood. Not all White people are enraptured by the White Spatial Imaginary. Some may grudgingly accept it. But even these "progressive" Whites typically benefit from it through the privileges and protections it has secured for their *neighborhoods*.[11] Similarly, many Black citizens may reject this way of thinking, but not all. Many are invested in the White Spatial Imaginary, especially if they have figured out how to thrive by "playing the game" of White spatial politics.[12] It should thus come as no surprise when Black politicians, business owners, or police officials push for an agenda that further exacerbates segregation.

Segregation is linked to police surveillance through the White Spatial Imaginary. When policing is organized in response to White racial fears, it becomes a driving force behind neighborhood segregation. As sociologist Monica Bell notes, "One of the basic strategies of police work is managing space, which includes surveilling people within physical communities of color and repelling people who, to police, seem out of place."[13] As I explore next, police surveillance in Baltimore fits this description. It has been about making the city "feel safe" for those living in the White L by monitoring and regulating the movement of Black bodies in White spaces.

The way the spy plane program was deployed echoes this history. Without understanding this past, you cannot understand how the program was structured by a kind of playbook for police experimentation that was built on segregation. The spy plane is just one more experiment in a long line of failed attempts at tech-driven crime control that, I speculate, may ultimately fuel some of the root causes of violent crime.

Making the City Safe for Commerce

The foundational surveillance experiment in Baltimore is the professional, uniformed "beat cop" himself. Beginning in the 1840s, cities across the United States began to introduce what was then a relatively untested idea: that police should be professionals rather than volunteers. Following what they saw in London, New York, Boston, Cincinnati, Philadelphia, and Detroit, a new political party in Baltimore, known as the American Party, aimed to professionalize the city's police with fancy blue uniforms, a regular salary, a centralized command structure based on military rank, and weapons.[14] Reformers thought this more formal (and armed) look would help police not only better react to crime but also function as a deterrent. When people saw these urban soldiers walking down the street, the thinking went, citizens would be less likely to even contemplate breaking the law. The fetishization of militarized policing as a form of deterrence was stamped into the form of the BPD from the outset.

Deploying them on March 1, 1857, political leaders had a central mission for these officers: to enforce a new set of ordinances that were designed to make the streets feel safe for commercial trade among Baltimore's White owning class, especially those who traded commodities in the buzzing port. What were these commercial traders selling? Among other things, enslaved people. Baltimore was a key node in the intra-US slave trade, and the port serviced slave ships.[15] Many of the new ordinances governing this commercial space were about movements and behaviors in public that city leaders saw as a "nuisance" to economic activity, such as standing and loitering, using foul language, or jostling and annoying passersby. Enforcing these ordinances required a huge expansion in resources and manpower. Hundreds of officers were hired to spread out over the city in scheduled patrols to conduct visual surveillance. The police budget ballooned from $70,000 in the early 1850s to almost $260,000 by 1860.[16]

Importantly, the police's mission was focused on deterrence. Property holders were encouraged to notify authorities even when they thought something criminal was *about* to occur. "By the 1860s," writes historian Adam Malka, "the city's leaders were authorizing policemen to protect property holders as their primary responsibility."[17] Officers were instructed to be responsive to property owners' sense of fear about crime and to proactively pacify that fear by patrolling these areas.[18]

Video Patrol

It's not difficult to draw a straight line from the origins of professional policing in Baltimore to the contemporary high-tech era. A good example, which deeply influenced how the spy plane operates, is the city's well-known CCTV system.

CitiWatch, as it is called today, had its origins in the early 1990s in the efforts of a Baltimore business association called the Downtown Partnership (an expansion of the Charles Street Management Corporation). Much like their nineteenth-century forebears, this group of business owners was concerned about the presence of panhandling and unhoused people, which, they worried, made downtown feel unsafe for consumers and tourists.[19] Their mission, in their words, was to "expand the cleanliness and vibrancy of downtown."[20] In 1990, they founded the Clean and Safe program, paid for by a surcharge on association members' property taxes. Among other things, the program funded a private security force, trained by BPD officers, to carry out foot patrols and "broken windows" policing downtown.[21]

In 1995, Clean and Safe program leaders heard about London's use of a municipal CCTV network to combat crime. They thought the technology might help downtown.[22] Because this technology had never been used in the city, the association was unsure if it would infringe on citizens' rights. Even though it wouldn't be owned by police, the system still had worrying implications for privacy. "The first thing the partnership did was hire a law firm," recalls Tom Yeager, the head of public safety for the Downtown Partnership and a former BPD executive.[23] The association secured a crucial judicial opinion about the privacy implications of CCTV. "The Fourth Amendment does not recognize the reasonable expectation of privacy and activity conducted in full view of public streets and sidewalks," the judge ruled. He continued, "Thus, the police may lawfully observe, with the naked eye, without first obtaining a search warrant, activity that

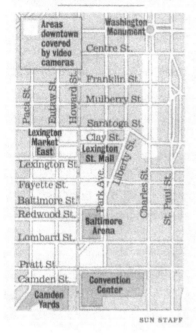

FIGURE 7. Location of original Video Patrol program, 2001. (Permission from Baltimore Sun Media. All Rights Reserved.)

is knowingly exposed to the public on municipal streets and sidewalks. The police may also capture these observations with a ... camera."[24] An officer looking at a camera recording of people on the streets is no different from him looking with the naked eye, the judge determined. CCTV had its legal footing.

The association chose a fascinating name for the program: Video Patrol. They won support from Baltimore's first Black mayor, Kurt Schmoke, to conduct a "pilot program" to see if the cameras would work. "I know that some people get nervous of Big Brother," Schmoke told the *Baltimore Sun* in 1996, "but I think everyone wants us to make Baltimore safe. This is a giant step toward doing that." Despite

Schmoke's emphasis on making Baltimore feel safe for "everyone," one gets the sense that this word was a stand-in for "consumers." Then BPD Commissioner Thomas Frazier said the quiet part loud when he gushed about the new system to reporters: "People can really feel safe when they come downtown to do business. We are doing what we can to provide a safe and prosperous commercial district."[25] The word *people* here means "shoppers." There's that White Spatial Imaginary at work.

The Partnership used a $40,000 federal JAG grant to purchase sixteen, fixed-point cameras that would record to VHS tapes.[26] The association could have put these cameras just about anywhere downtown, but they chose a specific location: a small slice of the neighborhood around Charles Street that borders the Black Butterfly.[27] This location is crucial to understand. As shown in figure 7, the area includes two of Baltimore's landmark commercial properties, which are a kind of barometer of the city's economic status in Baltimore's White Spatial Imaginary. Lexington Market, a large, covered pedestrian market, and Camden Yards, the city's celebrated Major League Baseball stadium, are part of an area that is a huge draw for tourists and wealthy suburbanites. If this essential class of visitors is scared off, the thinking went, Baltimore will be in trouble.

By the 1990s, citizens who traveled to the Black Butterfly's western wing to obtain drugs, panhandle, carry out petty theft, or even get drug treatment at the state's first methadone clinic had become a major concern for property owners downtown.[28] The main mission of Video Patrol, then, was to monitor this border between the edge of the White L and the edge of the Butterfly's western wing, which runs along Martin Luther King Jr. Boulevard.

The cameras were not monitored live. As shown in figure 8, they recorded to tapes that could be viewed in a remote "kiosk" by an off-duty or retired officer, but *only if they captured criminal evidence*. If there was nothing of interest on the tapes, they would be erased after ninety-six

FIGURE 8. A Video Patrol monitoring kiosk, 1996. (Permission from Baltimore Sun Media. All Rights Reserved.)

hours.[29] This emphasis on "recording" but not "monitoring" was often used to reassure the public that there was an added level of privacy protections. Yeager notes that, even though the judicial opinion the association won allowed them to conduct live monitoring and archive the footage, "We wanted more limitations than what the law allowed. . . . So, there were no dossiers, there was no storing of information."[30]

The cameras were fixed-point; they could only look in one direction. This meant the program's designers had to be strategic about where they pointed the cameras. Yeager recalls they put up cameras "in an area of West Baltimore that had a lot of vacant buildings, a lot of car break-ins, a lot of drug dealing, illegal vending, alcohol use, methadone clinics, and so forth." But they pointed the cameras specifically at parking lanes. As Yeager notes, "This was done primarily for a focus on preventing car thefts and breaking into cars, which downtown . . . is our biggest crime."[31]

Even though the cameras would be placed inside the White L, then, their purpose was to keep "outsiders" from the Butterfly under watch when they moved through White space. Much like nineteenth-century foot patrols, Video Patrol was meant to defend private property within commercial neighborhoods by watching those responsible for "nuisances" to trade who were infiltrating commercial space. This is further supported by what Yeager and other designers of Video Patrol do *not* talk about. There is little mention of murder or gun violence in any of the public discourse around Video Patrol, a noticeable absence given the homicide rate during the 1990s was the highest in the city's history.[32] Yeager admits that they never really considered getting community input on where the cameras might be put and what kind of crime they might be used to combat. "Community input? Not really," he recalls. "We went and put up cameras."[33] The cameras were meant to serve their primary clients—downtown business owners—so why would they consult anyone else?

The Video Patrol program quickly expanded from sixteen cameras around Lexington Market to over two hundred cameras across the entirety of downtown. Yeager recalls that this expansion was based on a positive internal evaluation of the pilot program conducted by the Downtown Partnership and the BPD. Yaeger claims the evaluation showed a "50-percent decrease in crime over the previous year" in areas around the cameras, as well as high approval ratings from "a market study of pedestrians" and "retailers" in the area.[34] The data from this evaluation are not public, however, and I could not locate them.

The *Baltimore Sun* also approved of the program, though with reservations. In a 1996 editorial, the paper cautioned,

> We hope everyone involved understands the limitations of these video officers. They can see all and do nothing. Effective policing still requires the city to have adequate numbers of human officers near

enough to respond to what they see through the cameras' eyes. But placing 200 cameras in downtown locations where they can see what occurs in public—not private—should give comfort to both Baltimoreans and visitors to our city. The possibility of being caught on videotape may deter some criminals. And the recorded evidence provided by the cameras could be enough to send other miscreants to jail.[35]

On the one hand, this editorial oozes the language of the White Spatial Imaginary. The point of the cameras is to generate "comfort" for (we can presume) wealthy White "visitors" and a deterrent fear for (we can presume) Black "miscreants." On the other hand, the comment is also a remarkably clear-eyed critique of tech boomerism. The editors make light of the fact that the program is an experiment run by human beings. The cameras are not magic. They cannot be all-seeing and cannot generate this comfort/fear dynamic without human labor. "The surveillance cameras, if used as stated," the editorial notes, "will simply be high-tech tools. The humans using them will determine their worth."[36] At the heart of this supposed panopticon are subjective human eyes. Still, the tone from the *Sun* was one of resigned support.

Shortly after the successful implementation of the Video Patrol program, a group of concerned businesspeople in Greektown, an historically Greek-immigrant neighborhood at the far eastern edge of Baltimore, began their own surveillance experiment in 1997. Hearing about the success of Video Patrol, the Greektown Community Development Corporation formed to, in the words of a representative, "keep Greektown from regressing and do something before it gets too late."[37] The language is a perfect example of the White Spatial Imaginary. "Regressing" means losing a grip on the unequal advantages that this (now) White (formerly non-White immigrant) community has made over the decades.[38] The group put up two cameras, monitored by workers they called "Citizens on Patrol." They placed the cameras strategically, almost like sentries at a gate, in a borderland between Greektown and the predominantly Black and Hispanic neighborhoods to its immediate west. The cameras were immediately used to target "nuisance" crimes. "Police have arrested people for loitering and urinating in public" by catching them on camera, reported a district commander in 1999.[39]

We thus see how the segregation of the eastern wing of the Black Butterfly gets reproduced technologically. With the Video Patrol program in the center of the city and the Greektown experiment on the eastern edge, commercial property owners were now able to more effectively monitor and repel the flow of Black and Brown bodies into White spaces. They were like two technological border walls set up around preexisting lines of segregation.[40]

Blue Light Cameras

In the early 2000s, police began to engage in rapid experimentation with different kinds of cameras, all of which turned out to be limited and glitchy. First, they tried "microwave cameras," which transmit a signal over the air. These proved to be far too expensive to operate and were quickly abandoned.

Next, they tried Portable Overt Digital Surveillance System cameras. These became known to locals as "blue light" cameras, because of an infamous blue light that strobed brightly when the camera was in operation.[41] These became a notorious eyesore. Not only were they unpopular with residents because of the annoying light, but they were also expensive (about $30,000 apiece), clumsy, and unreliable. Footage could only be viewed by physically going to the camera to access a hard drive inside the unit. Sometimes cameras were placed so high that the drive could only be reached with an electrician's truck.[42] They also broke down frequently. In 2006, for example, a newspaper reporter was shot while walking the street at night, and his assailants fled in a van that passed right by a blue light camera. When police went to pull the footage, though, they found it hadn't been recording for weeks.[43] The police commissioner once referred to the blue light camera system as an "albatross" and noted footage from it was only ever used in a handful of investigations.[44]

Despite a string of failures, Baltimore city officials would continue to push for an expansion of CCTV in the city. Once they found the best and most efficient cameras, their actions imply, the concept would finally start to bear fruit consistently.

CitiWatch

The 9/11 attacks were a turning point in the history of surveillance in the United States. With the ensuing Global War on Terror, federal money for surveillance technologies, funneled through the newly created Department of Homeland Security (DHS), flooded local police departments to beef up security.[45] Baltimore, with its small but significant port, received millions of dollars under this "anti-terror" regime and used it to create one of the largest CCTV programs in the world.

Between 2001 and 2004, the city spent upward of $25 million in DHS grants to build a massive public camera network that integrated with, and eventually swallowed, the private Video Patrol system.[46] This system had two key differences: it would be based in a centralized operations center where a paid analysis team would live monitor the cameras, rather than just view recorded footage; and it would archive the footage, rather than delete it after ninety-six hours. Whereas the early CCTV systems had always included the assurance that the police were "just recording, then deleting" and not "monitoring" or "archiving" pedestrian behavior, the mission had now crept right across that line.[47]

FIGURE 9. Inside the new Watch Center, 2005. (Permission from Baltimore Sun Media. All Rights Reserved.)

Though the BPD thought of CitiWatch as a mere outgrowth of Video Patrol, it was qualitatively different. Because the cameras were now live monitored, video analysts would need to act as the eyes for officers on the ground and quickly communicate important information about crime as it unfolded in real time. A brand new operations center in police headquarters, known as the Watch Center, was built to house this activity. It would cost upward of $600,000 a year to operate.[48] As seen in figure 9, the Watch Center was a far more technical form of analysis, which had never been tried in the city. Analysts weren't simply retrieving retrospective information from a tape. Now they would need seamless, coordinated communication between the camera feed, civilians staffing the center, and police.

In the lead-up to this expansion, local police and state officials focused on terrorism and used war as the main justification to deploy this relatively untested system. "We're at war," Dennis Schrader, director of Maryland Homeland Security, told reporters in June of 2004. "The purpose of the . . . system is to provide for the homeland defense . . . while also reducing crime and public disorder."[49] As Schrader indicates, in conjunction with warding off terrorism, CCTV could combat everyday "disorder" by deterring criminals and other "nuisances." Domestic crime and terrorism can and should be addressed by the same system.

The *Baltimore Sun* worried about what equating terrorism with street crime would invite. "The news that the city government plans to install and monitor 24-hour surveillance cameras . . . is just plain creepy," the editors wrote in the summer of 2004. Even though officials have said the system is needed to "protect East Coast transport, government, business and cultural gems from terrorists," they worry that the system "would later be used for other purposes. . . . Before authorities start watching Baltimoreans," they instruct, "they owe promises that public debate will precede any 'mission creep,' such as later using the system for police surveillance."[50] In retrospect, this editorial sounds naive. For one, this expanded system already represented a form of mission creep—from "just recording and deleting" to "monitoring and archiving." It also seems unfathomable now that CCTV would be used for anything but everyday policing. Of course the system is going to be used mainly for nonterrorism-related enforcement. This is evident in the way city officials looked to New York and London as role models, both of which used CCTV primarily for crime control. It goes to show just how powerfully the anti-terror hysteria of the post 9/11 atmosphere distracted mainstream journalists from the real intentions of the system.

War and terror were used to nudge the mission of surveillance across a couple of important lines: from recording to monitoring, and from looking at the White L to looking at the Black Butterfly. In the winter of 2004, the BPD publicly unveiled its plans for expanding

City to install closed-circuit TV cameras

Baltimore is adding three high-crime corridors to its patchwork plan to install closed-circuit TV cameras with 24-hour monitoring throughout the city to help deter crime. The three corridors will be added to the camera network that is up and running in the Inner Harbor and the planned expansion of the network into the west side of downtown and the Canton waterfront.

Three corridors for cameras:

1 **Park Heights Avenue:** 9% of Northwestern District's shootings and homicides
2 **Greenmount Avenue:** 10% of Eastern District's shootings and homicides
3 **E. Monument Street:** 25% of Southeastern District's shootings and homicides

Source: The City of Baltimore   SUN STAFF

FIGURE 10. Police announce CitiWatch expansion corridors to the public, 2004. (Permission from Baltimore Sun Media. All Rights Reserved.)

CitiWatch, which would place cameras in three new areas that cut deeper into the borderlands of the Black Butterfly: Park Heights Avenue, Greenmount Avenue, and Monument Street.[51]

As seen in figure 10, police threw off any pretense about terrorism in making this announcement to the public. The cameras' mission would be to control gun violence, specifically, through a deterrence effect. The expansion corridors were justified using a combination of arrest data and opinions from district commanders, targeting areas where shootings and homicides clustered—what police call a "hot spot."[52] The choice of these areas also mirrored a pattern of behavior that stretches back to the founding of the BPD. Each of these additional corridors sat in a racial borderland. They followed historic lines of segregation.

Adding a few dozen cameras at a time between 2004 and 2010, the system that would come to be called CitiWatch began to form a kind of technological border wall between the two Baltimores. If you look at the system today—now over one thousand cameras strong from over $5 million in sunk costs[53]—it takes on a familiar shape: the Butterfly and the L. As shown in map 4, there is a cluster of cameras in the downtown business district—the original system used to calm the nerves of anxious shoppers. The rest of the cameras cluster around the borders of the Westside and Eastside.

There are two types of neighborhoods, however, that I would describe as "camera deserts." Perhaps predictably, the wealthiest and Whitest residential areas of the city, in the northcentral and waterfront districts (the "body" and "foot" of the L) have few cameras. But also conspicuously absent are cameras in some of the Eastside and Westside areas that have the highest numbers of shootings. These camera deserts, which caused havoc for spy plane analysts, are the result of structural disinvestment during the experimentation phase of CitiWatch. Instead of putting cameras directly in West and East Baltimore, where shootings actually cluster, they are concentrated on the borderlands and commercial areas. It is in this sense that I think of CitiWatch as following the segregational logic of the White Spatial Imaginary. It's deployment pattern follows the mission of preventing violence from "spreading" into areas where it "shouldn't" be allowed to go.

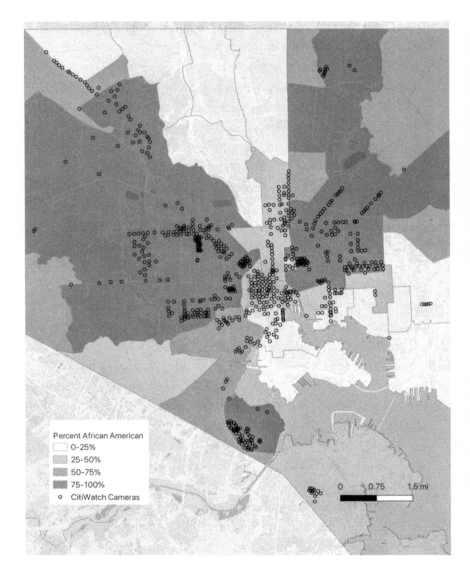

MAP 4. Racial map of CitiWatch cameras. (Data from Open Baltimore.)

Testing CitiWatch

Experimentation for the BPD often means "trying out"—that is, using a system without giving much thought to documenting effectiveness or preventing harms. CitiWatch exemplifies this kind of pseudoscientific approach. As it was promoted to the public, CitiWatch was ostensibly being deployed to deter *violent* crime, but there was then no data to indicate that CCTV would actually work on violent crime. The logic seems to have been that "crime is crime." If, as was reported about Video Patrol, cameras can deter property crime by solving more cases or scaring off opportunistic offenders, then the same logic must hold for gun crime. That's a big assumption. The truth is that despite the public rhetoric city officials had no idea if CitiWatch would deter murders and shootings in "hot spots." In 2005, after a reporter detailed a few studies from the United Kingdom that showed no significant decreases in violent crime from CCTV, Elliot Schlanger, the city's chief information officer, responded to the challenge by saying, "I've never seen a report that says CCTV *increases* crime."[54] Schlanger justified the expensive and experimental system with a kind of tongue-in-cheek comment that amounts to "well, at least it can't hurt, right?"

The lack of concern about demonstrating the effectiveness of CitiWatch continued in the period between 2005 and 2010, during which dozens of cameras were added without a systematic review. BPD executives can often be found talking to the public about how CCTV showed a "reduction in crime" in all the early test sites, but they never say what *type* of crime, say whether crime was reduced over the *whole* neighborhood or simply displaced, or provide examples of unanticipated *harms* that the cameras have caused, such as increases in police shootings or wrongful stops and arrests. These are important omissions. CitiWatch was being used in an entirely different way than prior camera systems, so, was it working? Police and city officials were never required to answer this basic question. They just pushed ahead blindly.

It wasn't until 2011, fifteen years after the first cameras had come, and six years into the CitiWatch experiment, that the BPD got around to independently evaluating CCTV. The program was part of a three-city study by the prestigious Urban Institute, which conducted an apples-to-apples, pre/post comparison between areas of the city with cameras and similar areas without to see if CCTV actually *causes* crime to decline.

The findings are stunning. The cameras have *no effect on violent crime* outside the downtown business district. Larceny, robbery, vehicle theft, and other nonviolent crime types decline in just about every place there is a camera (though not Northwest Baltimore), but not violent crime.[55]

Tellingly, the areas where the cameras were most effective at reducing nonviolent crime were in the borderlands between the Black Butterfly and White L (downtown and around Johns Hopkins University). Deep in the Black Butterfly, such as Westside neighborhoods beyond Martin Luther King Jr. Boulevard, even reductions in nonviolent crime were less substantial. In one camera-laden area around Mondawmin Mall, which would soon become the epicenter of the uprising over the killing of Freddie Gray, the technology showed no effect on crime, nonviolent or otherwise.[56]

Puzzlingly, the report's authors used these data to tout an optimistic future for CitiWatch. They note that "the above findings create a strong case for the positive impact that Baltimore's cameras have on crime." This conclusion is justified by a "crime is crime" mentality. Crime, in the abstract, does indeed decline because of the cameras, according to the data. But nowhere in the report does it state that, given the public rhetoric that CitiWatch would reduce record violence by deterring gun crime, the cameras seem to be doing nothing of the sort. That folks around Mondawmin Mall (they're shoppers too, right?) are being heavily watched but not receiving much benefit, even for reducing property crime, does not even register as a central finding.

The crime-is-crime mentality, and the political work it does, is even reflected in the way the study presents its statistical tables. The authors do not always break down crime trends by type, unless the analysis showed a statistical *decrease* in that type. So, while a decline in violent crime is reported in the downtown business district study area, where there was a significant effect, the *lack* of a significant decline in violent crime in all the other test sites is not included in the tables. The reader is thus subtly encouraged to see the cameras working everywhere, even though the effects are geographically (and racially) uneven. Once again, one gets the feeling of the White Spatial Imaginary at work here. If crime is going down in the commercial districts and the borderlands, alleviating fears among those in the White L, that's all we need to know. The tech can be said to "work."

However you wish to interpret the Urban Institute's findings, city officials ran with it as proof of concept to expand the system even further. Primarily under the reigns of Mayors Sheila Dixon and Stephanie Rawlings-Blake, both Black women, the CitiWatch system grew markedly between 2005 and 2015. In part, Dixon and Rawlings-Blake were responding to the popularity of the system among vocal city residents, especially older Black homeowners living in areas with a lot of gun violence and White homeowners living in the borderlands.[57]

One reason CitiWatch may have become so popular, even among residents who were not seeing huge benefits, is because of media coverage. The *Baltimore Sun* was one of the biggest voices to tout the success of CitiWatch. In a 2011 editorial covering the Urban Institute report, titled "Not Big Brother," the editors wrote,

> No one likes the thought of Big Brother constantly looking over their shoulder. . . . What the cameras do exceedingly well, however, is act

as a deterrent in much the same way a police officer posted on the corner might. In that sense, they're an extension of the eyes and ears of the policeman on the beat, and they're effective for exactly the same reason: People who do others harm are apt to think twice if they believe someone may be watching.[58]

The editorial feeds into what I've been calling the boomer/doomer hype cycle. By naming CitiWatch (Not) Big Brother, the editors imbue the system with greater powers of sight than it actually has. While this is usually spun into a "creepiness" critique—CitiWatch is "constantly looking over your shoulder"—it can just as easily be turned into a "necessary evil" framing. The editors make this turn precisely. "We may not like the idea of cameras watching over the city," they conclude, "but at a time when our ability to put more cops on the street is limited, they are an effective option." The editors reproduce the "solutionist" mentality of tech boomerism. They neglect to mention, however, that in one of the most important hot spots, West Baltimore, this "effective option" is having little to no effect on violent crime.

With positivity for CitiWatch stoked by the city's paper of record, it's no wonder the system became a fixture, even though it never really worked as the promised "solution" to gun crime. Residents were being fed imprecise information that ignored differences in effectiveness on specific *types* of crime in specific *neighborhoods*. In retrospect, not communicating this more nuanced picture was a seriously missed opportunity. In another universe, where the White Spatial Imaginary is not dominating politics, one can imagine police, city officials, and neighborhood representatives working together to find out *why* cameras don't seem to affect the violence around, say, Mondawmin Mall, and investing in a more neighborhood-specific plan. Some of those millions of dollars used to put cameras into the borderlands—laying cable, installing poles, cutting down trees, constantly repairing the pivot motors on the cameras, expanding the staffing of the Watch Center—could have been used in a more effective way.

A "Laboratory for Spy Tech"

After the hollow success of CitiWatch, there was a veritable surveillance tech boom in Baltimore between the late 2000s and early 2010s, mirroring the wider tech boom in Silicon Valley. This cycle of experimentation was driven by public-private partnerships between police and tech companies, a market in which PSS would become the latest player. As in other places, such as Los Angeles or Gaza, surveillance became privatized, and thus big business for tech firms hungry to demonstrate their tools.[59] The BPD was willing to try just about anything. By 2016, *Wired* magazine called Baltimore "America's laboratory for spy tech"—a city that "checks off all the requirements to build a modern American urban panopticon."[60] One thing didn't change from past cycles, however: experimentation almost always took place in the borderlands and featured untested and sometimes even dangerous technologies deployed with no oversight or public reporting.

Gunshot Detection

In late 2008, the BPD began to show an interest in gunshot detection technology, which had been tried in a handful of cities across the country since the late 1990s.[61] Using software that monitors microphones mounted on poles and roofs, gunshot detection systems are meant to decrease police response times by determining the location of a gunshot and sending it to police within seconds of the round being fired. Several companies at this time had flooded the market to try to secure lucrative contracts in high-crime cities like Baltimore.

Initially, the BPD was hesitant to try it because of the hefty price tag and "uneven results." "We think it's a promising technology," a mayor's spokesperson said, "but we're sort of waiting to have it perfected before the city considers making an investment."[62] They quickly found out they didn't have to wait. In November 2008, a large military contracting firm called Planning Systems Incorporated gave their experimental system, called SECURES, to the campus police of Johns Hopkins University *for free*. The BPD was happy, then, that Hopkins police would use their neighbors to the east as test subjects. This version of the technology had been developed for the US military using Defense Advanced Research Projects Agency (DARPA) funds and had been deployed in Iraq and Afghanistan to detect sniper fire. The company was seeking to "attract attention from college campuses across the country" to expand its market presence and chose Hopkins as a test site.[63] It installed ninety-three microphone sensors in the notoriously tense borderland between Hopkins (the northern branch of the White L) and the northeast district (the edge of the Butterfly's eastern wing).

Just a month later, the results were in: SECURES was a failure. The system had only detected two possible gunshots and both turned out to be false positives (one was a firecracker and another could not be sourced). Campus police noted that they even moved the sensors farther eastward into the Black Butterfly proper to see if they could get more positive hits, but nothing improved. Inexplicably, the director of Hopkins security remained impressed. "It's another layer of high technology," he told reporters, adding that city police should explore its use.[64]

Despite the failure, the BPD saw the Hopkins trial as an opening to launch their own experiment. Officials argued that the Hopkins experiment was a failure not because of the technology itself, but because it was deployed in an area that didn't see enough gun activity to warrant the system in the first place. "We told all these companies that if they put them up, we'd love live demos to work out the kinks and figure out what its capacity is," a mayor's office spokesperson told reporters in 2009.[65]

They chose to experiment in the borderlands. Installing sensors from several different companies along Monument Street—in East Baltimore, just north of the White L's "foot"—officials were immediately let down. They told reporters the systems had already "led to several false alarms." This process of experimentation lasted for more than five frustrating years, ending in nothing. In 2014, then BPD Commissioner Frederick Bealefeld reflected, "If I had to rate it on a scale of A through D, it would be a D-minus-minus. . . . It was a horrible, horrible failure."[66]

The BPD nonetheless *continued* to invest energy in gunshot detection. By this time, one company, ShotSpotter, had emerged as an industry leader. A highly critical evaluation of ShotSpotter's work by the Center for Investigative Reporting was released in 2016, showing an overall ineffectiveness at aiding in arrests for gun crime.[67] But the BPD was undeterred by these data. In 2016, they secured an $860,000 grant from Bloomberg Philanthropies to pay for ShotSpotter service in East and West Baltimore. After two years of testing, in which police placed the sensors in locations that would "avoid the false positives from loud noises echoing among the downtown buildings," they publicly unveiled the system in the summer of 2018.[68]

The department renewed their ShotSpotter contract in 2021, to the tune of $760,000, despite yet another negative evaluation of the technology's accuracy and effectiveness from Chicago's inspector general.[69] As this evaluation notes, there is now enough data from other cities to suggest that ShotSpotter alerts may more often be false positives than real positives.[70]

Unfortunately, ShotSpotter has been unwilling to let experts analyze the false positives problem with their own internal data. Because of the for-profit relationship between this company and the BPD,

ShotSpotter data is not public. We therefore cannot know if ShotSpotter is actually improving anything about violent crime in Baltimore. Moreover, we have no idea what kinds of new risks or harms the technology may have created, such as sending amped-up police officers to respond to a false positive, looking for a person to violently pacify.[71] Other than the Monument Street corridor, we don't even know where ShotSpotter sensors are located. In fact, BPD officials told me that *they* don't even know the locations of the sensors. That is proprietary information that only ShotSpotter knows.[72]

With even more gusto than CitiWatch, then, ShotSpotter came to Baltimore through a series of public-private partnerships. One of the factors driving this form of experimentation seems to be a "let the money guide you" mentality. Rather than reach out to companies to tell them what they need, the BPD is flooded with offers, even freebies, to use the city for beta testing. Baltimore is a juicy proving ground. The BPD thus becomes more like a passive consumer than an active provider of public safety.

Stingray

The privatization of surveillance makes surveillance practices less transparent.[73] Whereas the CitiWatch era featured publicly available plans and maps of camera expansions, those kinds of details are rarely given in the era of public-private partnerships. Sometimes, as with the 2016 run of the spy plane, the public is not even told that new surveillance technologies exist. One tool used in Baltimore—the cell site simulator, often called a stingray—is a notorious example.

A stingray is a small box-shaped device, developed for the military, that mimics a cell phone tower. It forces a target's phone to connect and thus reveal the phone's location and other metadata. As a surveillance tool, it allows police to use a cell phone's location as a proxy for the location of the person who owns the phone. If this sounds to you like the vehicle location analysis conducted with the spy plane, you're right, the two technologies are cousins. Both use data to identify a person, even prior to being connected to a crime, by tracking the locations of objects associated with them (phones in this case, cars in the case of the spy plane). In the process, they expose hundreds of bystanders to mass location tracking simply because they are in the area of a target.

The BPD acquired a stingray for something like $200,000 from Harris Technologies as early as 2007 (it is unclear exactly how it was paid for).[74] Nobody in the public, press, or even the city's own judicial system knew about it. Detectives were using it in arrests and then showing up to testify in court, but they would not reveal how they had acquired a suspect's location or that they even had such a technology, a tactic known as "parallel construction."[75] Why all the secrecy? In order to get the technology, the BPD signed a nondisclosure agreement with the FBI to never discuss its existence or how it was used.[76] It was only in 2014, during the routine discovery process of a murder-for-hire case, that federal prosecutors learned the gadget had been secretly available to police for years.[77]

The justification for the secrecy, detailed in the FBI's nondisclosure document, strikes a familiar tone. Federal officials told the BPD they should drop any case involving a stingray when defense attorneys or judges inquired because not doing so "would reveal sensitive technological capabilities possessed by the law enforcement community" and thus "result in the FBI's inability to protect the public from terrorism."[78] Just like with CCTV, the fight against terror was justification for a new surveillance technology to slip into investigations of "other" criminal activity by local police.

In 2015, *USA Today* discovered exactly how that "other," nonterrorism-related use had played out. The BPD hadn't just been using stingray occasionally. It had gone on a veritable cell phone tracking binge, using the device over 4,300 times between 2007 and 2014, dwarfing the frequency of use in other agencies.[79] What were they using it for? Not anti-terrorism. The majority of cases were run-of-the-mill crime cases, like armed robbery and theft.[80] In one case, police tracked the phone of a woman who had stolen a credit card to pay for two months' rent on a storage unit. In another case, they tracked the phone of a woman who had been sending "threatening and annoying texts" to another person.

What is more, barely half the cases ended in a prosecution, with about a third of them dismissed outright.[81] The BPD was using stingray to churn out small-time arrests, only for the technology to fall flat in court. In 2015, the State's Attorney's Office was forced to go back and review every case involving a stingray to determine whether or not to toss it out—a massive undertaking that gummed up the courts for months.[82] By using the technology hastily, without first figuring out if it would stand up in court, police ultimately thwarted their own mission to "send a message" of deterrence to citizens. If anything, the message was "catch and release."

Stingrays also generate new harms. They can disrupt the functioning of a cell phone, preventing a user from being able to call 911. More importantly, the bluntness of a stingray search can expose an entire neighborhood to this harm. When you turn them on, they will capture information about anyone in the area by forcing all neighboring cell phones through them. They are thus not only prone to mistakenly sweeping innocent bystanders into investigations, but also disrupting the ability for bystanders to use their phones in an emergency—a clear threat to public safety.[83] This is not to mention the threats to privacy from the mass gathering of metadata. Again, if this sounds familiar, it is a risk shared by the spy plane's wide-angle gaze.

In 2017, ten years after the stingray had spread to many local police agencies, the Superior Court of DC ruled that police need to secure a warrant before using it.[84] Today, the stingray is seen as a highly invasive and imperfect technology that should only be deployed in rare circumstances with clearly articulated reasons for suspicion.[85] Throughout the period of experimentation, however, when police used the technology for just about anything involving a cell phone, those risks were just being discovered. The FBI and Harris Technologies gave the BPD the power of experimentation, and the agency abused it by "trying it out" however they wanted, without accountability. When it didn't work in court, disrupting the system, they didn't have to report that. When it stopped a bystander's urgent 911 call from connecting, something that must surely have happened at least once, police didn't have to disclose that. These are significant harms from experimentation that emerged without oversight.

One important question to ask, then, is *where* were those risks experienced? Who bore the costs of the technology's harms until the problems were discovered? A coalition of civil rights groups, assisted by legal scholar Laura Moy, sought to find out. Mapping every time the BPD used a stingray, and laying each use over a racial map of the city, they uncovered a familiar shape.[86] Each of the dots in map 5 represents an instance of stingray use. They cluster in the Black Butterfly to a shocking degree, especially given that White L neighborhoods see the same kinds of small-time theft and nuisance behavior on which the technology was commonly used. This map shows how the risks of this untested technology were unevenly distributed. Each stingray use not only disrupted the cell phones of *target* suspects, but also *bystanders* in the same neighborhood. How many 911 calls in the Butterfly were blocked by this experiment? How many people from these neighborhoods sat even longer in jail, awaiting trial because of cases tainted by the stingray? We will never know because the department did not keep track.

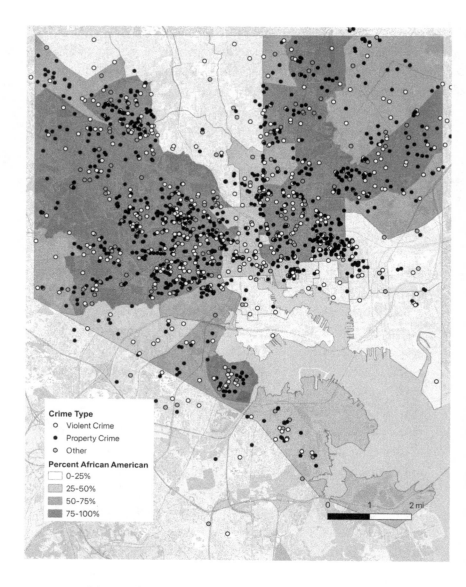

MAP 5. Racial map of stingray use by Baltimore police, 2007–2014. (Data from *USA Today*.)

The Power of Experimentation

Taken together, the histories of CCTV, gunshot detection, and cell site simulators bring a troubling pattern of BPD behavior into relief. First, the agency treats the city like a pseudo-laboratory. It is a laboratory without any of the safeguards, transparency, and public notification of risks that one associates with real experimentation.[87] Baltimore is only a "laboratory for spy tech," then, in the sense that police and city officials are "trying stuff out," not in the sense that they are actually looking to see what works through some sort of rigorous scientific process. The BPD seems unfazed by the idea of deploying technologies that have a completely mysterious risk profile, but then also makes little effort to detail these risks to citizens. During the era of public-private partnerships, this lack of transparency has become even worse. It is often only the private tech companies, who profit from the technology's adoption, that know about these risks.

Second, the technologies the BPD experiments with often turn out to have major problems. They are more limited than advertised, glitchy, and ineffective, and can even cause new problems that limit the police's ability to deliver cases that will end in conviction. Sometimes, as with the stingray, they even directly threaten public safety. The justification for deploying these pseudoscientific tools is almost always to deter *violent* crime specifically. When the technology proves ineffective at doing this, however, the BPD seems perfectly willing to push forward. One gets the sense that, as the department is flooded with new gadgets that continually disappoint, police feel pressure to keep using them anyway. The city, or some grant agency that the city wants to impress, spent a lot of money on it, so, "we might as well use it."

What becomes crystal clear from this history is how the problem of experimentation, supercharged by privatization, gears together with the White Spatial Imaginary and the police's role as agents of segregation.⁸⁸ The thread running through the history of police surveillance in Baltimore is this: police wield experimental surveillance technology, often at a profit for outside companies, in an effort to make the city "feel safe" for those in the White L by creating a deterrent fear in the Black Butterfly. This helps reproduce geographical racial lines of division from generation to generation. Surveillance technologies are deployed like a technological wall in the already established racial borderlands.⁸⁹ These walls help secure the economic vitality of White L neighborhoods while disproportionately exposing Black Baltimoreans to unknown risks. When there is fallout from these risks, as we saw with the stingray debacle, it is Black folks who typically bear the brunt.

If it is indeed true that segregation is a "forming cause" of violent crime, then this history should raise one crucial worry above all. It is not far-fetched to speculate that experimentation with pseudoscientific technologies might actually *contribute* to the root causes of gun violence in hyper-segregated cities like Baltimore by further demarcating racial borderlands and concentrating unknown risks in already destabilized race-class subjugated neighborhoods. Could these untested tools indirectly *fuel* gun violence? That's not a question I can answer here, but perhaps it is one we should be asking.⁹⁰

The Playbook

When the spy plane came back to town in 2020, then, this is the historical fabric into which it was woven. The spy plane was deployed using a well-established playbook that looks something like this:

1) Just try it out. Pilot programs give police license to engage in risky experimentation without oversight or accountability. If the technology is limited, triggers false positives, unleashes new harms, or destabilizes the court system, nobody will be held to

account, so just "try it out." If police like it, then they can evaluate it later to see if it "works."
2) The Black Butterfly is a petri dish. As long as a surveillance technology shows no blowback on White neighborhoods, it is okay to use the Butterfly for experimentation. Especially if the tech is effective at reducing "crime," in the aggregate, in borderlands around the White L, it doesn't matter too much if the technology is ineffective in Black neighborhoods or makes mistakes that affect Black lives.
3) Don't say what you're doing. First, deploy it to "see if it works" and then "unveil" it to the public after. If you do tell the public early on, only tell them what they want to hear and obfuscate or hide the real risks, especially around privacy. Terrorism is a good cover. Another is to limit the technology's uses initially in order to "protect privacy" and then allow it to "creep" into more invasive uses later, when no one is paying attention anymore.
4) Let the money guide you. If a company or grantee comes to the city offering a freebie, take it, even if you don't have the organizational sophistication to use the technology properly. Grantees and tech companies can and should tell police what they need, not the other way around.

As I will show in the next few chapters, this playbook is precisely what guided the BPD's deployment and experimentation with the spy plane.

5   Big Brother's Bigger Brother

For more than a decade, a small group of plainclothes Baltimore police officers, known as the Gun Trace Task Force (GTTF), ran a racketeering scheme from inside the department. Their celebrated position as an elite unit that "got guns off the streets" allowed them to routinely falsely arrest, intentionally harm, and even steal from those they believed "deserved it," such as people with past drug and gun convictions.[1] On a snowy January day in 2016, they falsely arrested a man named Jawan Richards, perhaps to steal money or drugs from him, and, in the course of events, shot him through the windshield of his vehicle. Richards survived, but the officers were able to put him in prison by saying Richards tried to run them over with his car. This wasn't the first time they had said things like this. Unlike most days the GTTF worked, though, that day there was a secret spy plane filming them from the sky. The officers, like everyone else in 2016, had no idea the surveillance plane even existed.

The footage of this incident is blurry and unremarkable, but it captured a few tiny details that would later confirm the GTTF's pattern of behavior. The arrest report filed by the officers says, "Mr. Richards attempted to flee the scene, rapidly reversing his vehicle into the unmarked police vehicle located at the rear of his vehicle, causing detectives . . . to open fire towards the vehicle, striking him once in the neck."[2] The GTTF officers claimed to be acting in self-defense. They also claimed to have found an illegal firearm in the car. PSS footage of the incident, however, hints at a different story. In the company's summary, analysts note, "We observed vehicle behavior [that] matches BPD story of vehicle backing into officers, but not into *officers'* vehicle. Officers' vehicle was *in front of* suspect car."[3] In other words, Richards backs up suddenly, but the layout of the cars in the WAMI footage doesn't match officers' sworn statement. Judging by the aerial imagery, it seems highly unlikely Richards could have threatened the officers' lives with his car.

Other details contradicted what police said, too. The officers claim there was a long verbal altercation with Richards before they shot him, but the WAMI footage shows just fractions of a second between officers arriving on the scene and Richards's attempt to flee. It's not a full debunking, but something does look fishy.

There was more. Analysts also saw the detectives meet just prior to the shooting a few blocks away from where Richards's vehicle was parked. Within seconds of Richards's vehicle beginning to move, the detectives, in remarkably coordinated fashion, move to intercept. As federal investigators would later reveal, this was typical of the GTTF's tactics.[4] They often placed illegal GPS trackers on the vehicles of their targets, allowing them to precisely coordinate their attacks. Without knowing it at the time, then, PSS had some of the first corroborating evidence of perhaps the most corrupt police squad in the agency's history.

It would take four years for this footage to surface. As I discuss more below, the imagery eventually got to Richards's defense attorney, Ivan Bates, who would later become the State's Attorney for Baltimore City. Using PSS's analysis, Bates convinced a judge to throw out Richards's conviction.⁵ Richards would later collect $850,000 in damages from the city.⁶

Though it is often only thought of as a tool to watch citizens, the spy plane can also watch police. In sociology, this is known as *sousveillance*, or "watching from below."⁷ The word *below* refers to the power hierarchy between the state (police, prisons, courts, government and military officials, etc.) and citizens. Where the state is typically on top, in terms of power, and therefore watches from "above" (*sur*veillance), it has also long been the case that citizens have watched them back, using the power of collective action from "below" (*sous*veillance). From the Black Panthers' citizen foot patrols in the 1960s, to the infamous Rodney King video in 1991, to the citizen cell phone footage that sparked the George Floyd uprising in 2020, sousveillance imagery is now a major part of the public conversation about the criminal legal system.⁸ Can visual technologies, normally tools of race-class subjugation, also be used for resistance and reform?

Between 2016 and 2020, when the spy plane program returned to Baltimore, Ross McNutt discovered that there was a great deal of public interest in using the spy plane to watch police. During these four long years, I followed closely as PSS went through what McNutt would call a "rebranding." He and other supporters held seventy-six community meetings, primarily in West Baltimore, to show citizens what the spy plane could do and ask for their feedback and support. McNutt wanted to hear what residents hoped for in the spy plane, should it ever return to the city. What he found out was that people wanted community control over the program so they could better hold police accountable. Could the spy plane actually break from the toxic history of experimentation that preceded it and become a tool of Black empowerment? Could the communities most heavily watched by "Big Brother" use the very same technology to become the state's "Bigger Brother"?

Turning the Cameras Around

Simmons Memorial Baptist Church sits near the apex of a triangular-shaped intersection in the neighborhood of Sandtown, known as Penn North, one of the historic centers of Black Baltimore. Duane Simmons, the charismatic pastor who heads the church, first heard about the spy plane like most everyone else: through its sudden exposure in the media in 2016. "When I heard it for the first time," he reflects, "I was skeptical. Because our community has been hurt. [Police] have used those items that should be used with good intent, but they used them for sinister purposes. You know, Big Brother in the sky, and all that."

A year after the program had been paused, however, Simmons was approached by McNutt. He asked the pastor if he would be interested in joining a "focus group" to learn more about how the program worked and to provide feedback to PSS. Simmons, still skeptical, called on Joyous Jones, one of his most trusted community members, to go to the meeting to find out more. Joyous is a highly respected church elder who has tirelessly worked for decades to serve Sandtown through church-based "wraparound services," as she often calls them: food, shelter, medicine, education and job opportunities. She also worked for a prominent defense attorney who has represented victims of police misconduct. Compact, white-haired, and with a gentle voice, Joyous's calm demeanor belies an incredible energy to fight for what she believes will save her community. So, she answered her pastor's request.

On a hot summer day in 2017, Joyous sat down at a long table inside the spy plane operations center across from McNutt and took out a pen and paper, ready to write down everything he said.[9] Next to her were four younger people, one of whom was Archie Williams. Sporting long dreadlocks, a wide headband keeping them out of his eyes, and a bright blue Superman T-shirt, Archie looked uncomfortable being there. After glancing at Joyous, Archie returned his eyes to his phone, almost refusing to give his attention to McNutt. He had been invited to the meeting by one of the young parishioners of Memorial Baptist. Unlike Joyous, Archie, in his early thirties, had spent much of his life in and out of jail. Up until just a few years ago, he was living on the street, only able to pull himself out by unexpectedly obtaining Section 8 housing. He reinvented himself, in part, by starting a small nonprofit aimed at combatting homelessness called Community With Solutions.

Joyous and Archie didn't know each other before they came to this meeting. This makes sense because, in many ways, they occupied different sides of some of the great social divides in Black Baltimore and many other African American communities—old and young, women and men, working class and unhoused, church and street.[10] Though outsiders often treat Black Baltimore as monolithic, there are complex tensions that shape its politics. Perhaps unexpectedly, the spy plane bridged these divides for Joyous and Archie. Though they both came to the meeting as harsh skeptics, by the time they left, they would join up to help McNutt "rebrand" the program and bring it back to Baltimore.

In interviews years later, Joyous and Archie tell me that this meeting was like a conversion experience. They both point to exactly the same moment when their opinion of the spy plane turned from skepticism to excitement. After McNutt had spent nearly an hour going through a long PowerPoint deck, describing how the spy plane could "solve otherwise unsolvable crimes," he noted that the plane could be a work-around for one of the core barriers to closing cases: a lack of trust in police. "When we often ask investigators, what do they have," McNutt said, "their typical response is, 'I got nothing.' And in part that's because people don't feel comfortable enough with the police to come forward, becau—." Joyous cut him off short, saying,

"Because a lot of the police here are part of the problem." She told McNutt about "the one-hundred-and-sixty-three-page report" from the DOJ documenting long-standing racial abuses.[11] She detailed a 2017 incident in which a narcotics officer was caught planting drugs in order to frame a suspect, but tried to stage a scene using his body cam footage to make it look like the drugs were "discovered."[12] Though he didn't speak up at this point, Archie nodded his head in agreement, the first time he had engaged in the meeting.

McNutt heard Joyous out and had a ready response. "One of the nice things about what we do is—because we're only providing information as to what happened at the scene, we do see the responding officers. . . . We can tell when the responding officer showed up and what they were doing beforehand." He then pulled up a slide titled "Potential Police Misconduct" and detailed one case when PSS analysts caught police lying. In 2016, PSS helped a defense attorney gather evidence that contradicted an officer's sworn statement that a man was selling drugs from his house. The footage was enough to convince a judge to throw out the charges. "Magically, the case went away the next day," McNutt said proudly. "So that is something we could do."

Archie's body language changed at this moment, his back straightening and eyes widening. "So, this is for, like, everyone?" he asked McNutt excitedly. "This is not a law enforcement tool. This not somebody else's 'I'm gonna keep my eye on you.' This is a community assistance, 'I gotta eye on the city. Period?'" McNutt nodded enthusiastically. "What I'm asking you is," Archie said, "you're not on no one's side but your own? That's what I'm asking you." "It is just video imagery of the whole area," McNutt replied matter-of-factly. Archie was hooked.

It's only later when I interviewed him that I found out just what this moment meant to Archie. "My baby brother was a victim of police brutality," he told me.

> My brother died at the hands of police officers. That's why I can never be one. So this [program] is an opportunity for me to police the police and also have documentation to the people as a service, which is—we don't have that. There's no kind of service right now in this city that provides that, unless you hire some private investigator. . . . This is at civilians' fingertips.

For Archie, the spy plane had shifted from being a "partisan thing" to being a community service.

Excitedly, Archie reiterated to McNutt just how groundbreaking this was, explaining, "This is the concept you have in Baltimore with people right now. . . . The trust is gone. So by you coming with this program, it's like, 'We are [just] eyes. We are not looking for nobody per se, but we are overseeing things.'" From the corner of the room, Carday, a young man who brought his toddler to the meeting and who had said nothing up to this point, chimed in, "We're big brother's bigger brother." Everyone laughed.

Joyous would take this information about PSS's sousveillance program back to Pastor Simmons, who was equally impressed. He would later coin the phrase "Turn the Cameras Around" as a clever tagline to promote the program to community members. In 2019, for example, on a talk radio program popular with Black listeners, Simmons preached,

> The premise of Turning the Cameras Around, all that means is that we are going to encourage the community by telling them that you are not the only ones being watched. It's extremely important that, if you want to gain the confidence of the community, we make sure that those who are behind the camera have skin in the game. Skin in the game means that we are not just watching you, but we are watching ourselves. Turn those cameras around. Because the sinister use of them would be minimized considerably when an individual

behind the camera is as much a part of it as the person in front of the camera.

Simmons knows the "sinister" history of surveillance technology in Baltimore, but sees McNutt's program as different because police have something to lose if they misbehave in front of the cameras. Already you see the distance between police and community has slightly shrunk, such that the "we" in his speech is ambiguous. With the spy plane, he seems to be saying, it is sort of like everyone is policing everyone.[13] Joyous, Archie, and Pastor Simmons see utopian possibilities in this. Where a dystopian, "doomer" critic might see the plane's persistent gaze as creepy, Black community members saw the exact opposite—a technological "solution" that could create transparency and address distrust in police. PSS was happy to play up this angle to community members to gain their support.

Community Buyback

That the spy plane could adjudicate the broken relationship between police and Black Baltimoreans was just one of the reasons community members were drawn to McNutt's pitch. They also thought the program could help deal with another set of social divisions: the notoriously tense relationship between "street" and "church" within the Black community itself.[14] Spy plane supporters wanted the program to help police more effectively differentiate between the very small number of "violent repeat offenders" who are living high-risk lifestyles in the streets, and who are responsible for most of the violence, and the vast majority of residents who are not involved in crime but who continually have run-ins with police because of imprecise and racist tactics.[15] As made clear by the Department of Justice, in its 2016 investigation of the BPD, the agency has long engaged in unconstitutional stops in Black neighborhoods, the vast majority of which result in no charges or convictions.¹⁶ Because of this kind of imprecise, dragnet surveillance, community members like Archie, Joyous, and others who became supporters experience both too much policing and not enough: too much unfocused, arbitrary, and unreasonable police contact, and not enough focused, effective, and justified police work. They want "more" policing, if by more we mean more precise—policing that aggressively pursues and punishes only those few who threaten the safety of everybody else.¹⁷

It is on this point of effective and precise public safety that the spy plane exposed a deep political rift within Black Baltimore. Black activists, many of whom were younger and who harbored a deep skepticism about policing, characterized the program as naive and dystopian. They thought the spy plane could never deliver on its promises.¹⁸ In keeping with the prison abolition movement, they wanted the BPD to shrink to make way for alternative solutions that really could provide the targeted public safety they needed.¹⁹ In fact, it is these young activists who came up with the phrase "spy plane" in the first place.

These criticisms grated supporters like Joyous, especially when they came from educated folks who did not live in her neighborhood, such as legal "talking heads" from the ACLU or Black activists from local universities. These critics, many of whom were claiming to speak in the best interests of poor Black people, only emboldened her. "I find it difficult for people to tell me what I need," she told me.

> We have Black-on-Black crime, okay, but this idea that we don't need any more policing, we need something else? It's not working. You know, a good ol' slap on the wrist? I'm sorry . . . that's not working. You know, we have to put these violent offenders away, and then maybe save the children. . . . You've got to get the career criminals off the street. And we have a tool to do that [in the spy plane]. Let's do that. And *then* let's help the ones that are coming out. We have wrap-around services for them.

Joyous sees critics' calls for shrinking the footprint of policing as replacing real justice with a "slap on the wrist." She sees it as not only ineffective, but also disrespectful of her local knowledge. Tough, aggressive policing has its place, she argues, when it is actually targeting the right people. Historically, it has not. For her, then, the spy plane might finally deliver the kind of precision she had been craving, but which the BPD had never been able to achieve. Even better, this increased precision would free up the energy and resources wasted on unnecessarily broad surveillance, such as stop-and-frisk patrolling, to be put into "wraparound services" for young people who, she emphasizes, still have something about them worth saving. She wants precise, aggressive surveillance for the few and improved wraparound services for the many. Perhaps ironically, then, the way the spy plane could fulfill Joyous's vision of targeted precision would be to watch the whole city, police and all.

Listening to the debate within Black Baltimore about the spy plane changed McNutt. He began to think of PSS as more than just a technology service provider. What if the company could not only help punish the "bad guys" but also help uplift the "good guys" in the community by becoming a source of jobs and career development? Hosting more focus group meetings between 2017 and 2019, Joyous, Archie, and McNutt talked to dozens of other West Baltimore residents, many of whom were affiliated with other churches, proposing a radical idea: what if PSS hired Black residents and trained them to become analysts for the company? If the spy plane were to ever fly again, McNutt suggested, the program could help put money and jobs back into the community by enfranchising its members in the surveillance industry itself. In a way, it would be "the people watching the people."

In early 2020, just prior to the program launch, McNutt spoke alongside Archie to a largely Black audience who followed Archie's radio show.

MCNUTT: What we do is take young people, mainly people who can play video games, and we train them up to be . . . investigative analysts. We are looking to hire twenty to thirty analysts locally. . . . Our analysts make about fifteen dollars an hour, starting out, and go up to twenty. . . . Usually people who work for us get their [security] clearances through us and then go off to work with the DOD. We provide a stepping stone to those careers.

ARCHIE: Wow, that is an economic change for real. And we need that. . . . So for someone who is just coming home [from being incarcerated]—it's gonna take about two to three years to get this gigantic gorilla called a felony off his back. Would you be . . . willing to hire an ex-felon?

MCNUTT: Well, we are going to work with expungements . . . to make sure that returning citizens, those who have paid their debt to society, would be eligible. Some of those details are still being worked out . . . but we're pushing as hard as we can to make sure that the job is open to as many people as possible.

Having spent many hours talking with West Baltimoreans about the need for good jobs in fields like technology, McNutt saw an opportunity to use the spy plane as a way to uplift young people. Pushed largely by Archie, he even entertained the idea of helping the formerly incarcerated expunge their criminal records and become eligible to work as contractors. In a distant future, McNutt could see himself guiding these young people into careers in the federal intelligence industry, where they could make upward of six figures.

For Joyous, it was this utopian dream of using the surveillance program to get "buyback" into the community (by both watching police and investing in young people) that ultimately hooked her on the program. "You have the community watching the community," she explained to me a few weeks before the first plane launched. "That was my winning sale."

PSS was tapping into an energy that is widespread in the criminal legal system reform movement today. Across the country, cities are experimenting with new forms of public safety that involve hiring and training civilians from high-crime neighborhoods, often men with criminal records and gang affiliations, to engage in violence reduction. Violence interrupter programs, to take a notable example, of which Baltimore has been an early test site, take men with credibility on the street and train them to intervene in neighborhood disagreements and gang rivalries before they result in violence. Though PSS personnel didn't think of themselves as this kind of alternative to traditional law enforcement, the company's "jobs program" seemed to play into this trend by emphasizing direct community input and labor.

Though it is difficult to gauge widespread public opinion about the spy plane program, what little data we have suggests that PSS's rebranding effort was successful. There's evidence to show that aerial surveillance was popular with Baltimore residents, and especially so among Black residents. A survey conducted by the University of Baltimore midway through the program's six-month trial showed that just over 55 percent of all respondents and over 66 percent of Black respondents supported the program.[20] Support was even stronger among older citizens, retirees, women, those without an associate's degree, and those living in high-crime neighborhoods, ranging as high as 75 percent approval.

More intriguing, though, is a distinct racial gap about privacy worries. Among those surveyed, White respondents, even those living in high-crime neighborhoods, were more likely to complain that the spy plane violated their privacy, compared to Black respondents. This racial gap is more than a little ironic, given how the program, in practice, experimented primarily on Black lives. It also shows how

Black residents, at least those surveyed, put privacy below other concerns when considering the spy plane. Privacy advocates' repeated warnings about invasiveness, then, seem to have fallen on deaf ears for a large proportion of Black Baltimoreans (though certainly not all).

In sum, Joyous, Archie, and other supporters were clearly convinced that the spy plane could become a voice for and defender of the community. How did this vision match up with the reality of how the program played out?

Police Are Hard to Watch

Police are not like us. As historians Brittany Arsiniega and Matthew Guariglia note, police in the United States have historically been treated as "supercitizens"—receiving all kinds of special treatment that shield them from punishments and offer privileges other citizens don't enjoy.[21] One of these special treatments is a kind of "superprivacy"—a set of protections that prevents the public from scrutinizing officers' behavior and seeking accountability for corruption and misconduct. One mechanism of superprivacy is a culture of "reciprocal silence" within the criminal legal system. As part of an agreement among brothers in the fight against crime, cops (and often prosecutors) don't snitch on other cops, or look the other way when they see misconduct.[22]

Reciprocal silence becomes a major cultural barrier to the success of misconduct investigations when the tools of these investigations are controlled by police. Body cams, to take a well-known example, tend to give police a lot of discretion in how they record and who can access the footage.[23] Police can simply "forget" to turn on their cameras or "accidentally" turn them off. Even when misconduct is captured by body cams, it is often police executives who get to choose how to release the footage, thus allowing police to better control the narrative. Though video of police behavior would seem to be objective, then, the way the cameras are deployed and the footage managed is often not.[24] Though PSS promised the spy plane could watch police, given officers' superprivacy, would the program actually be able to hold police accountable in practice?

Reconsidering the GTTF Case

At first, the Jawan Richards case, which begins this chapter, suggests that, indeed, the spy plane can work well as a police accountability tool. The story sounds like this: GTTF officers wrongfully stopped and shot Richards. The plane saw it. The officers were held accountable. Richards's conviction was erased, and he received compensation. What more do you need to know? A lot, it turns out.

The only reason Richards's defense attorney, Ivan Bates, came to know about the spy plane footage was because of an independent investigation by a reporter named Brandon Soderberg. Soderberg had written a number of meticulously reported stories about the program, including PSS's attempt to rebrand itself as a sousveillance tool.[25] Soderberg told me that in the fall of 2018, while talking to McNutt on the phone for one such story, McNutt said, almost out of the blue, "Well, you know, back in 2016 my plane captured a police shooting." Soderberg was intrigued. "Ross is a bit of a salesman," he recalls, "so I thought he was maybe not entirely telling the truth." McNutt couldn't remember the name or date of the case, but gave Soderberg a 2016 flight log. Soderberg did some digging on police shootings that year and figured out that the plane had been up when Richards was shot. He met with McNutt to view the footage and then passed the information on to Ivan Bates's team, who then reached out to McNutt directly.

I obtained a recording of a video chat from 2019 between McNutt and Ivan Bates's law clerk. During this call, McNutt makes a stunning revelation: both the Baltimore Police Department and the Carroll County State's Attorney's Office, which had taken on the case due to a conflict of interest, tried to bury the spy plane footage. And even when they failed to make the evidence go away, they tried to overturn PSS's interpretation of the imagery by misleading Richards's attorneys.

Police and prosecutors had the spy plane footage the whole time Richards was being tried, purposefully withheld it for months, then purposefully mislead the defense about what was in the footage, in order to protect the conviction. Richards ended up sitting in prison for three years on a plea deal while GTTF officers continued to falsely arrest and rob other citizens. This is superprivacy at work.

According to McNutt, PSS analysts immediately reached out to someone high up in the BPD in January 2016 as soon as they realized they had captured the police shooting. "We put together our briefing," McNutt told Bates's clerk. "We told [our liaison inside the BPD] we had seen this event, and he went over there to meet with the police, to have them come by and look at it. And what we got told was, essentially, 'Don't look at it.'" Police sat on the information for nearly eight months, during which time Richards was forced to take a guilty plea. In September 2016, he was sentenced to several years in prison for second-degree assault and possession of an illegal firearm. Richards's attorney, of course, was oblivious to the existence of the footage and to the existence of the spy plane at all, just like everyone else back then.

Part of the reason the BPD seems to have sat on the case, McNutt surmises, is that the program was still secret. The notion that this dystopian-sounding program might have evidence on something as sensitive as a police shooting, especially in the wake of the 2015 uprising against police brutality following the death of Freddie Grey, would surely have been terrifying to the BPD in early 2016. Pursuing a misconduct case in this charged environment and having to do it with evidence from a secret spy plane? No thanks.

Once the program came to light in August 2016, however, McNutt says the BPD returned to the case as part of a review of all the investigations PSS had done during the short-lived trial. "Once it became public," McNutt recalls, "this was the *one* case they raised a big stink about. Out of everything we did, *this* was the one case that caught their attention." The case was forwarded to the Carroll County State's Attorney's Office in order to avoid a suspected conflict of interest between one of the GTTF officers in the case and a Baltimore prosecutor. Officials from Carroll County came to the operations center to examine the briefing and, guided by McNutt, the spy plane footage itself. McNutt's recollection of this meeting to Bates's assistant is revealing. He reflects,

> Well, I will tell you, the Carroll County prosecutor—I don't know if I should say this out loud, but he was an asshole to us. . . . His major fear was putting his guilty plea at risk. He was pissed because *we* put *his* guilty plea at risk, and he said that explicitly.

McNutt recalls being shocked at this response, saying, "It was almost like, he didn't care one iota about the case. He was almost like, 'You're gonna make me lose this game.'" This response didn't surprise Bates's law clerk. "Spoken like a true State's Attorney," he sighed. "We see it all the time in this city, you know: 'By any means necessary.' Get the stats, get the convictions, pad the numbers and, you know, go about your day. It doesn't seem to matter what really happened or how it happened." The threat of "losing" a conviction was scarier to prosecutors than failing to find corrupt cops.

What happened next, then, is a textbook example of superprivacy. The State's Attorney of Carroll County, just a few days after

Richards had pled guilty, having just reviewed the PSS footage, having just complained to McNutt that it threatened his conviction . . . sat on the evidence, just like the BPD had done when the incident first happened. But he went even further. A full month later, in October 2016, he sent a letter to Bates's office. The letter said,

> I . . . reviewed images of this incident captured by PSS and the wide area surveillance program, and, based on our review, there is no information captured by surveillance that is inconsistent with any of the police investigation or civilian interviews. PSS's report *incorrectly* concludes that the footage captured by the plane does *not* show . . . Richards struck the police vehicle.[26]

Carroll County officials, hardly experts in WAMI analysis, told Bates that the imagery had been misinterpreted by PSS. Providing their own interpretation of the data, and playing it off as more authoritative than McNutt's, they concluded the imagery confirmed everything police said about the incident. They essentially told Bates's team that the footage was irrelevant.

According to Bates's clerk, prosecutors misled things even more. They gave the impression that the few fuzzy still shots of the crime scene included in PSS's report was all that was available. "We didn't know that [prosecutors] ever saw a video," the clerk recalls. There was no mention that PSS had captured hundreds of sequential images, both before and after the incident. Bates's team didn't even know they could request the same sort of bespoke reanalysis of the footage that prosecutors got. "Had we looked at [this] and been able to talk to you at the time," the clerk tells McNutt, "things could have been *very* different." He suspects that, with the full video, Richards would never have pled guilty. "I would have loved to have briefed you back then," McNutt responds. "I'm sorry about that. And I'm sorry for your client, also."

The Problem of Interpretation

The spy plane is supposed to be able to hold police accountable. How can it do that, however, if police can so easily make the imagery go away or produce an alternate interpretation? Spy plane footage, it turns out, is far more susceptible to the power dynamics inside the BPD and the courts than one would like. Bates's clerk was alarmed at this prospect, telling McNutt,

> If BPD can just come back and say, "That's not what happened, that's inconsistent with how the police reported it," you know, "we're not going to take the surveillance plane seriously," then BPD is always going to be able to override anything that you ever submit. So then, you know, the surveillance plane kind of becomes useless.

Despite the utopian hope that the spy plane can become an independent witness, which can see guilty police as easily as guilty citizens, the Richards case suggests that police's superprivacy makes this very difficult to achieve in practice. Breaking through this requires a level of public trust, objectivity, and organizational transparency that are in very short supply in Baltimore (and, apparently, Carroll County, too). Bates's clerk reflects,

> We want this surveillance stuff to be its own stand-alone entity. But we need to make sure we treat it as such. We can't just let anybody make assumptions about what they see, you know. We got to stick with, you know, the *creator* of this system, what he says *he* saw.

Perhaps ironically, for Bates's team, the spy plane footage could only be treated as objective if PSS's *interpretation* of the imagery was privileged as the most authoritative. But McNutt and the analysts at PSS are not police. They are not supercitizens. So, when legal authorities wanted to overrule the view from McNutt's eyes, it was as easy as a prosecutor writing a misleading letter to a defense attorney.

The Richards case raises a crucial question about video evidence in police misconduct investigations. Why was it so easy for Carroll County prosecutors to create their own alternate view of the imagery, passing it off as authoritative without fear of pushback? Partly, it is about police as supercitizens, but I don't think that's the whole story. I would suggest there is also something intrinsic to the look of the imagery itself at play here. Something aesthetic is going on. As I explore more in the next chapter, spy plane imagery lends itself to multiple interpretations because of its abstract, grainy, and jumpy qualities.[27]

Throughout McNutt's conversation with Bates's clerk, for example, he continually describes and apologizes for the "blurriness" of the imagery. "This is right on the edge of our imagery, which is probably the worst shot on the worst day," he says sheepishly. He discusses how, because it was a snowy day, just after a historic blizzard, the glare of the snow on the ground "caused havoc on our gain control on our aperture," making it really hard to see dark objects, like cars, moving on the ground. He talks about how the incident occurred between rows of three-story buildings, preventing him from seeing the whole street. "It's like looking into a canyon, in essence," he says. At one point, the wheel of the plane comically peeks into the imagery for a few frames, as McNutt is running through the footage, blocking the view of the street where Richards and GTTF officers are approaching each other. By the end of his presentation to Bates's clerk, then, McNutt tries to tame expectations. "I don't know that our data would be a kind of instant replay kind of thing," he confesses. "I don't know that [it] would be adequate to overturn anything that an officer swore that he saw." Was it suspicious that authorities tried to suppress and then mislead about the imagery? Yes. But does the imagery itself actually show that police lied? Not definitively.

Given how challenging it is to interpret the footage, it is no wonder police and prosecutors could so easily devise an alternate reading to McNutt's. There are just so many ways one could reassemble what the herky-jerky movements of these blobs and pixels means. Prosecutors must have felt perfectly within their right to say that PSS misinterpreted the footage. As with police body cam video, we may want spy plane footage to act as a "mechanical witness"—raw data produced by a scientific machine that needs no interpretation—but in practice it is just as vulnerable to interpretation as any other kind of narrative evidence, perhaps even more so because of its graininess.[28]

What seems to have ultimately erased Richards's conviction (which happened three years after the incident, after he had already done prison time and been released) was not the spy plane, but the dogged persistence of Soderberg, the investigative journalist who alerted Bates. The fact that the evidence went missing, was misleadingly delivered, and finally got to Bates's team in such a circuitous way—not to mention that it captured imagery of (by then) *known* corrupt cops—was more than enough to get the conviction thrown out. It almost didn't matter what the footage actually showed. And besides, it turns out McNutt himself wasn't completely sure.

The Richards case suggests that the sousveillance function of the spy plane, just like its surveillance function, was riddled with uncertainties around accuracy, reliability, and transparency because of the toxic organizational environment of the Maryland criminal legal system. It was an experiment, deployed within a highly dysfunctional system, and no one knew if it would work. Above all, then, it shows that without an organizational structure that allows sousveillance data to be gathered independently of police and their lawyers, so-called "cop-watching" technologies are unlikely to actually be able to hold police accountable.[29] As I discuss in chapter 8, without putting the power to monitor police squarely in the hands of citizens, sousveillance may be more hype than reality.

Sousveillance in 2020

These problems continued in 2020. In one of his first announcements about the AIR program, BPD Commissioner Michael Harrison said the plane would *not* be used to watch police. Confused and shocked, McNutt pushed back and ultimately got consent to use the plane for sousveillance, but only in cases of police-involved shootings. The program could not be used, for example, to track police vehicle locations to document overtime abuses, a notorious form of routine corruption in the city.[30] With this rather delimited mission, McNutt set up a separate team of three analysts dedicated solely to sousveillance. No police shootings occurred during the 2020 trial period, however, and, to my knowledge, the sousveillance team was never used.

In addition to the rather narrow framing of the sousveillance service, I was also struck by the lack of engagement between PSS and defense attorneys. The extent of the outreach was a letter to the Baltimore Bar Association, sent after the program had already ended, saying analysts were available to consult for the defense. No defense attorneys used this service. Why, after all the talk around the program supporting "both sides" of the courtroom, was there so little effort to woo defense council? Partly it seems to be because of resistance from both the BPD and prosecutors. In June, while the planes were in the sky, PSS organized a meeting over video chat to explain the program to prosecutors and take questions (a privilege that wasn't granted to the public defender's office or other defense attorney networks). According to internal documents I obtained, during that meeting, prosecutors asked "a lot of questions" about whether the footage could be used by the defense. McNutt informed them that PSS had "set aside funds and analyst hours to support some defense requests, but that it was up to BPD to determine when and under what circumstances the data could be used." Rather than treating the sousveillance program as independent, then, PSS had given the BPD control over its shape. PSS were waiting for the BPD's blessing and direction.

Talking to people inside Baltimore's public defender's office corroborates this view. No one inside the office can remember PSS explaining how analysts could assist. Though McNutt had spent years promising Joyous, Archie, and other supporters that the plane would be a powerful new tool for defenders, when the rubber met the road in 2020, the sousveillance program never got going. So much for becoming Big Brother's bigger brother.

Promises, Promises

Sousveillance was not the only or even most important promise PSS made to the program's West Baltimore boosters. Creating jobs that would go to residents of high-crime neighborhoods was a big part of the "buyback," as Joyous called it, that got people excited. Joyous herself became a beneficiary of this promise. McNutt hired her as an HR and office manager, providing enough consistent work to allow her to quit her old job with the defense attorney.

I talked to Joyous in March 2020, just a few weeks before the first planes were scheduled to take flight, as she was still getting her office set up inside the operations center. As HR manager, one of her jobs was to help hire an entirely new team of analysts who, it was assumed, would be recruited primarily from her network of young people connected to Simmons Memorial Baptist. Without consulting her, however, PSS outsourced the hiring to a computer skills training organization that, Joyous learned, was owned by the son of a major politician that supported the spy plane. Though she didn't describe it in these words, I got the impression that, to Joyous, it felt like a political kickback to a person more powerful than her.

The computer skills organization ended up hiring twenty analysts, ten Black residents of West Baltimore and ten White women from an Orthodox Jewish community in North Baltimore. Joyous, in her characteristic way, was outwardly serene but livid on the inside. "This is supposed to be a buyback into *our* community," she told me in hushed tones in the days before the program began. "I have some people that have excellent criminal records. No crimes. Never been arrested." But before she even knew to put these names forward, the hiring had already taken place, cutting out many of the young people she had promised jobs. Thus, residents of the White L were given a piece of the company's much anticipated "jobs program" that was equal to what was given to residents of the Black Butterfly.

Though Joyous was displeased with how the hiring played out, the office did end up with a good number of Black analysts from high-crime neighborhoods. Many of them told me they were deeply proud to take part in the program. This pride stemmed mainly from their personal connection to the neighborhoods they were analyzing from the sky. Those grainy, pixelated streets, sidewalks, and buildings might look abstract to an outsider, but to them, they were connected to embodied experiences. Damon, for example, a twenty-two-year-old who grew up in West Baltimore, frequently worked cases right around his house. He remarked,

> I literally, like, live next to where a couple of these shootings happened. One case, there was a lady who was getting some chicken and got shot by some random idiot. Later that day, I went into that same place to get some chicken, right? I mean, I know these places and I live in these places.

Several Black analysts had lost loved ones to gun violence and told me that they could see how the plane might have been able to help them, had it existed earlier. Sharee, in her early sixties, lost her son in a daytime shooting years ago and worked for PSS with her daughter in the hopes that they could help other families of victims. She told me,

> I wish they had had the [aerial] cameras in 2011. Maybe we could find out who killed my son. It was nine thirty in the morning. He was at the store and [*claps hands to mimic the sound of a gunshot*]. They had a [CCTV] camera on it, but I don't think it worked.

Sharee could imagine a different path for her life, in which the spy plane overcomes the technical malfunctions that prevented a sense of closure about her son's death.

Beyond this personal mission, Sharee told me she was surprised at the sense of pride that came to her as someone working a job that traditionally excludes people like her. She recalls working a case in which her team saw something and called detectives in to look at the footage, which quickly led to a court order to make an arrest. "That's a big accomplishment for us," she said. "We felt like we were really contributing. And detectives seem, like, really appreciative. They really trust our judgment, more or less." There was a sense among Black analysts, then, that the program enfranchised them by putting them in the position of expert on gun violence in their own neighborhood. Though PSS didn't invest in the Black Butterfly to precisely the extent promised, there was at least some "buyback," and Black analysts relished the chance to contribute.

The pride that came with working for PSS, however, did not come without huge risks—risks that analysts were not warned about. In the first month of the program, analysts were told that only McNutt would ever have to testify in court, should one of the spy plane cases make it to trial. This was not true. Managers learned that, just like with any other forensic tool, attorneys would call on analysts to give an account of their work in front of a jury.

In late June (nearly three months into the job), managers gathered analysts to tell them that, indeed, they would have to testify, "even if you just touched a case." While White analysts found this worrisome because of the controversial nature of the program, most of the Black analysts panicked out of fear for their lives. Taking the stand might label them a snitch in the neighborhood, exposing them to retributive violence. "Are we gonna have some kind of protection? Will we have to tell our real name on the stand?" Ethan, a young West Baltimore resident, asked supervisors. John, a White manager from Dayton, replied, "That is part of what it means to do this kind of work. You are protecting your community, and in doing that, you open yourself up to the risks that come with testifying." He suggested that "it might be a good idea to get some protection." He suggested analysts start bringing a weapon with them to work. There were head shakes and sighs at this comment. Ejaife, cracking a half smile, asked, "This is going to sound strange, but, like, would it be okay if I wear like a fake nose and a wig or something?" The whole office exploded in laughter.

The light-hearted joking during this meeting hid an underlying fear that did have a significant impact on Black analysts' state of mind. Two analysts from West Baltimore quit shortly after this meeting, though I could never determine if they left because of the risk of testifying. The risk remained on the minds of those who stayed. Analysts began to make sure to change out of their company-branded shirts at the end of a shift before they walked out on the street or onto a public bus. Damon, having worked cases in which he had literally seen his neighbors on CCTV, reflected,

> [Baltimore] is like the birth place of "stop snitchin'." So that kind of stuff is real, and I do worry about [testifying]. But I just keep close, you know. The only person I'm talking to is like family members and close friends. I don't really tell them anything at all. . . . And they are kind of worried for me because, well, people are crazy. . . . But, overall they know that I'll probably be fine because I don't associate myself with a lot of people.

Living in West Baltimore while working for the spy plane meant that Damon was both the subject and object of experimental surveillance. This created a kind of chilling effect in his life, leading him to keep a lower profile than usual. While White analysts also worried about testifying, they would always be able to retreat into the White L and thus remove themselves from the immediate threat of retribution. Much like how the risks of experimentation were born unequally by the residents of the Black Butterfly in general, even inside the program the risks were shouldered unevenly by Black workers.

A Dream Deferred

The dream that the spy plane would become "Big Brother's bigger brother," by both watching police and enfranchising Black citizens in the crime fight, ran into both familiar roadblocks and unexpected risks. The main takeaway is this: actually watching the police in such a way that it would change the system requires much more than putting a spy plane in the sky.[31] The way the Jawan Richards case played out is a perfect example of just how difficult it is to watch officers in a way that would hold them accountable (even ones that are literally running an organized crime ring from inside a police agency). The lesson, then, is to not invest so heavily in technologies that purport to watch police without also investing heavily in *organizational structures* that will actually let the evidence gathered by these technologies operate independently of police.[32]

It's the same with using the citizens of high-crime neighborhoods as criminal legal workers. As I discuss more in chapter 8, there is an incredible amount of hope and hype around citizen-centered alternatives to traditional police experiments, like violence interruption. But we cannot forget that these programs are *also* experimental, putting already vulnerable citizens in extremely vulnerable situations, often for very little pay and with little psychological support.[33] Baltimore's justifiably celebrated Safe Streets violence interruption program, for example, lost two of its most experienced members to murder around the time of the spy plane experiment.[34] While nothing so extreme happened to PSS's Black analysts, they deeply felt the unique risks that come from "the community watching the community." They had to tailor their lives around these risks in ways that made them shrink from the public. It is imperative, then, that city officials acknowledge and honor these risks when asking community members to do violence work. In the simplest terms, that means better pay, but also that systems are put in place to support civilian violence workers emotionally with the types of risks, such as retribution for snitching, that police don't typically have to worry about.

In the end, there were a lot of broken promises for the Black citizens who backed the spy plane. When it was shuttered at the end of the six-month trial, there was a brief flurry of good news that the city of St. Louis would sign on with PSS in the winter of 2021. McNutt promised to keep as many Baltimore analysts as he could, working St. Louis cases remotely. He tried to keep the dream alive, but data from the Baltimore trial did not convincingly show an impact on crime.[35] Baltimore elected a new mayor who vocally came out against the spy plane as a "gimmick."[36] Moreover, as I discuss in the next chapter, the program became embroiled in a civil suit against the BPD for violating the Fourth Amendment. This quashed contract negotiations with St. Louis. Hope was crushed yet again. The experiment had failed, and all the Baltimore-based workers would have to be let go.

The only person who was kept on was Joyous, who, in her genius way, began using PSS's old office space to run computer training classes for residents of her neighborhood. Joyous organized a final farewell party for the analysts in the late fall. Reflecting on this moment, she told me how "you could see the sadness." They were losing more than just a job, she said. As one analyst said in a rousing speech, "I respect each and every one of you. This feels like a family to me." "It was a sight to behold," Joyous sighs. "You shoulda seen everybody hugging."

6 *Privacy and the Time Machine Problem*

"I'm gonna be totally honest with you," I tell Sarah and Ryan as we sit down to an interview in the spy plane operations center. "I find some things disturbing about this case in terms of privacy." They seem somewhat taken aback by what I've said, looking back with slightly confused frowns. "A car is not a person," I say, and PSS has been tracking cars that are associated with crime scenes, treating them as proxies for persons of interest. Then, going further afield, they have been surveilling addresses associated with those cars (proxies of proxies) going back days before the incident. In this case, these associations led them to surveil a suspect's mother's house for several days, and that just isn't sitting right with me. Sarah responds, "I see how you could say that, but even without this technology you often put an APB out on a car and try to find the person. How is this any different?" She adds,

> We know the car. We have the plates. So it's not like we're just looking at any old car around the mother's house. We are looking for *that* particular car. And that's just how we work. We look for things and track anything that's in the area and eventually we narrow it down to only the things that matter. Everything else gets tagged DNI [determined not involved], and that's the end of that.

Ryan agrees, adding that, yes, they are looking around the area and other people are getting swept up in the tracks, "but if you're not doing anything wrong, you have nothing to worry about." It's not like they are going to just look at a location indiscriminately, they reassure me.

As we're talking, we become aware of a shadowy figure hovering just outside the briefing room door. The door has frosted glass, and because of the light shining in from outside we can clearly make out the shape of Henry, another analyst, with his ear to the door. Because of the asymmetry of light, he obviously can't see that we can see him. Ryan yells, "Hey, we can see you! Do you need to come in?" But Henry doesn't hear because of all the noise in the operations center, so he walks away. "Okay, now that creeps me out," Sarah says. "I didn't know it looked like that from this side." Ryan asks, "Well what do you think about one-way mirrors?" "I like them when I'm on the other side of the mirror," Sarah laughs.

State surveillance is often organized according to a principle that philosopher Michel Foucault called panopticism. In a panoptic situation, the watched know that they're being looked at, but not exactly when or exactly how, and that makes them behave.[1] Like Sarah, people often describe this feeling as creepy.

When you're on the watcher's side of this asymmetrical exchange of gazes, however, this kind of creepy staring can seem like a benevolent thing to do. It is easy to disconnect from the chilling feelings of being watched because you trust and believe in yourself as a benevolent watcher. You know that you would never watch for the wrong reasons or watch people who don't "deserve" it. How funny, then, that Sarah and Ryan, in the act of telling me "if you aren't doing anything wrong, you have nothing to worry about"—probably the most commonly used defense of mass state surveillance—became instantly creeped out by being watched, even though they weren't doing anything wrong. What is more, they even knew that Henry's attempt to gain knowledge of their actions wasn't really panoptic. He couldn't see or hear much through the frosted glass, but they were still unsettled by his attempt. Somewhere deep down, intuition told them that it is better not to be on the side of the watched, even, and perhaps especially, if the watcher is not operating at full capacity. Who would want to be inside a one-way mirrored room, subjected to someone's half-cocked experiments, especially when the view from the other side of the mirror turns out to be a blurry mess?

The case I referred to in my conversation with Sarah and Ryan was a litmus test for the spy plane program's privacy policy. A homicide investigation that involved surveilling a suspect's mother's house based on data from a GPS tracker inside the suspect's car (provided not by police but by the Acura Corporation), this case expanded the scope of the program beyond what the public had been told PSS would be doing. The way the BPD handled this case opened the door to uses of the plane that not only violated the Memorandum of Understanding (MOU) the department had publicized to Baltimore citizens, but ultimately ran afoul of a new set of Fourth Amendment legal decisions that are upending the world of privacy law.

The activities in this case also raise an issue that, I argue, has been sorely overlooked by privacy advocates because of how "boomer/doomer hype" has so powerfully shaped the conversation: What if experimental surveillance technologies don't work the way police say they do? If they are more hype than reality, what are the implications for debates about privacy, which often presume that these tools are almost magically invasive and panoptic?

One way to shed light on this question is to look at how and why the spy plane program got itself into hot water around privacy when it had tried so hard not to. How did PSS's attempt to create a program with strict privacy protocols end up becoming what the Fourth Circuit Court of Appeals would call "a 21st century general search"?[2]

Move Fast and Break Things

Prior to the program's launch, PSS had forcefully argued that it would operate within a set of tight boundaries intended to protect citizens' privacy. One of the central features of this program was the design of the cameras themselves. PSS engineered the camera system to represent a person as no larger than one pixel in size. This "one pixel per person" resolution was created in order to maximize the coverage area of each plane's surveillance orbit, while still giving analysts enough visual information to track effectively. It also had, according to McNutt, built-in privacy benefits. "All we see is dots," I often overheard McNutt explain to the program's skeptics. The system "cannot see what you look like. . . . It cannot tell if you are man, woman, or child. It can't tell if you're Black, White, or any other color." It merely records people's "public movements," he stressed, as they travel about outdoors. When members of the public inevitably asked what happens when PSS cameras get more powerful, McNutt would reply that he would never seek to increase the resolution of his imagery. Instead, he would expand the breadth of the camera's gaze in order to allow analysts to track over even longer distances. "I would want to see wider," McNutt explained to me when I asked about improvements to the cameras. "Because what does a few more pixels [of detail] on a person add to the tracking? Not much." But expanding the breadth of the camera's gaze? That could mean the difference between knowing where a suspect is hiding out and losing him off the edge of the imagery.

When it came time to promote the spy plane program to the public in 2020, Commissioner Harrison parroted McNutt's language. Throughout his public presentations, he said the program was completely within the bounds of privacy law because it "can't see into anyone's homes" and only looks at "public places," which are already being heavily surveilled by police and private cameras. He told citizens,

We already have closed-circuit cameras across the city, and there are more coming. We have license plate readers that read license plates as cars go by, and there are more coming. There are all kinds of technologies that we have, and more are coming. This does not lead to a police state. As a matter of fact, people are being videoed more by each other than [the plane] will ever do. But there is no expectation of privacy in public, so there is no police state in public.

Harrison compared the spy plane to a big CCTV camera in order to place it in a less creepy-feeling context, with an established legal foundation. Drawing on several Supreme Court decisions, the BPD's legal team took the view that WAMI is essentially no different than any other kind of aerial surveillance, such as a helicopter.[3] In this way of thinking, the spy plane could invade people's privacy only if it could look inside their homes, which it certainly could not. In the open air, they would argue, people don't expect privacy anyway. As I explore below, however, this was a rather narrow explanation of what turned out to be an extremely complex and volatile legal landscape for privacy.

PSS also promised the public that analysts would never look at the city willy-nilly on behalf of police. Every time an analyst examined the imagery, they said, it would be tied to a known crime scene. Thus, surveillance would always be focused and tight, only sweeping into the investigation those who were known to be involved with a specific incident. "Analysts use imagery data to locate crimes, track individuals and vehicles *from a crime scene*, and extract information to assist BPD," states the MOU, signed in March 2020 and presented to city government.[4] "Tracks of individuals *to and from crime scenes* form the basis of analysis," it says elsewhere in the document.[5] Even earlier, in the years-long lead-up to the launch of the program, when PSS engaged in extensive consultation with Black community leaders, the company emphasized that the surveillance would be bounded. In one community meeting, hosted in 2017, McNutt assuaged worries about privacy by saying,

> Internally—and we put it on contract [in] our privacy program—we go to a crime that has been reported. . . . And we also put it on contract with the city that if a police officer says, "Hey, can you look here?," we can say, "No, we're not allowed to, because of the contract."

The company promised that the spy plane would only look at known crime scenes, and what is more, it would operate somewhat independently of police. PSS could therefore push back when police wanted to use the technology in ways that would expand beyond these tight boundaries.

It's important to see that both of these policies approach the issue of privacy through the lens of "creepiness." They try to make people feel more comfortable with the fact that police will be looking at them from the air. Rather than focus on the quality of the data and how it will be used—for example, how the company will be sure that the "pixels" it is tracking actually correspond to the correct person, or what the company will do with the data once an investigation is over—these policies aim to assuage people's visceral recoil from the idea of an all-seeing eye. This is one of the dangers of thinking about privacy with a "theory of creepiness"—all surveillance companies need to do to address privacy concerns is make the technology *seem* less creepy. That can function as cover for uses that may in fact be deeply invasive.[6]

In May of 2020, just a few weeks into full operations, PSS and the BPD broke many of the promises it made around privacy. Why? Because of a combination of investigatory exigency and a subtle but powerful commercial pressure to demonstrate the technology's effectiveness. As a private tech start-up, PSS had to "move fast and break things" in order to push the envelope and demonstrate to police, the largest consumers in the market, that they could generate results. One of the things they ended up breaking was the Fourth Amendment.

More Powerful Cameras

As promised, PSS maintained the one-pixel-per-person resolution to its camera system throughout the trial period. However, many detectives were visibly disappointed when they saw the spy plane imagery for the first time. One detective I met literally said, "That's it?" when he came to view the imagery on an impromptu visit. On several occasions detectives asked if the company could improve the resolution or, more often, add other camera systems alongside WAMI that could see individuals close-up. One detective told me that he was disappointed with the footage and that, during his tour of duty in Iraq years before, he recalls seeing persistent surveillance imagery that was much better.

McNutt seems to have felt the pressure to answer these worries about the imagery quality. He told several detectives that PSS was working on a more powerful system that would more than double the pixel power of the cameras to four hundred megapixels. This would allow them to see even more of Baltimore at a slightly better resolution, not enough to identify someone by their physical features, but more impressive nonetheless. It might even allow them to track at night without the use of night vision lenses. PSS planned to put this system into use in Baltimore in 2020 on a third plane, but was unable to finish and deploy it before the city terminated the contract.

Another detective, after having been briefed on a case, asked PSS, "What other kinds of toys do you guys have?" John, an ex-law enforcement officer with extensive military experience, headed up communications with detectives. He was happy to oblige these kinds of questions. He pulled the detective into his office to show him test footage the company made while bidding for a military contract. It was from a Wescam, which had been placed alongside the spy plane's wide-angle system. This camera, manufactured by L3Harris and used extensively in the US drone program, has the ability to "tell you the color of a guy's shoes" from miles away, John told the detective excitedly. The detective marveled at the quality of the footage and asked if the company could use this in Baltimore. McNutt explained that this would violate the MOU, but "maybe if we stick around [in the city], that's something we could add."

Though I never saw the company go back on its promise about the power of the cameras, then, it is clear that they were receptive to renegotiating these terms if their contract were to be extended. Moreover, these exchanges demonstrated just how desperately the company wanted to please police by putting a product in front of them that would look more impressive, even if that product was experimental and would violate promises around privacy. As discussed in chapter 4, the early privacy boundaries of a surveillance experiment in Baltimore are often just performative—they provide cover for the program to creep into a more invasive mission later, when the public isn't looking anymore.

Only Crime Scenes

The graininess of the imagery was one problem detectives had with the system, but not the primary one. Very quickly, investigators found that using the planes to only surveil bounded crime scenes was limiting because of the spottiness of the coverage. If a crime happened at night or during a cloudy day, or sat just outside the area of the plane's orbit, PSS would continually have to tell detectives that they couldn't help because the incident location was out of coverage. Again, wanting to make a good impression in order to preserve their foothold in the city, PSS felt the pressure to adapt. Rather than telling investigators no, if the crime scene didn't lie in coverage, as McNutt had promised, supervisors began to respond, "How else might we be able to help?" Rather than operating as an "independent witness" within tight privacy boundaries, PSS was becoming more like a kind of "help desk" for detectives.

By late May, just a few weeks into full operation, the boundaries that were set up around focused watching were run right over. Detectives began requesting surveillance into times and locations outside the actual incident, which they called "supplemental requests." They exploited a loophole in the MOU that allowed police to request assistance for matters not related to a specific crime scene, so long as the police commissioner himself had approved it.

In the MOU, supplemental requests are meant for support in "extraordinary and exigent circumstances," such as a kidnapping or train derailment.[7] In practice, the commissioner used it more broadly to allow surveillance in everyday crime cases. Through a supplemental request, signed off by the commissioner, police could surveil any person, place, or thing associated with a known suspect and (presumably, though I did not see this happen) any associations of those associations. In total, the program fully investigated sixteen supplemental requests (out of 186 investigations) and received more requests than that, which they didn't investigate due to lack of imagery or other limitations.

Not overly impressed by the spy plane, detectives clearly put pressure on PSS to "move fast and break things," opening up conversations about experimental cameras and, ultimately, a whole new style of "supplementary" surveillance that contradicted promises in the MOU. The Hutchinson Street case, which opens this chapter, was one of the first of these supplemental cases. By delving deeply into this investigation, we can see how the company's orientation toward "rapid prototyping" in a high-stakes, profit-driven environment became a threat to citizen's privacy rights.

The Hutchinson Street Case

The Hutchinson Street homicide occurred at night, which, right away, would rule out the use of the spy plane. But detectives found a different way in. They obtained two photographs of a white Acura that was believed to be the suspect's getaway car. This allowed them to pull both the license plate number and VIN and get the identity of the suspect (presuming, of course, that it was the owner of the car that was driving it at the crime scene). Pulling up more information revealed that the car was equipped with the AcuraLink system, an app that allows the driver to control features of the car using a smartphone. Like a lot of the smart technologies in the "internet of things," it also gathers location data using GPS. For consumers, this is meant to be a convenience feature. "Never forget where you parked by getting your vehicle location right in the app," Acura enthusiastically advertises on its website. Little did the suspect know that police could exploit this customer convenience feature for state surveillance.

Through the Acura Corporation, detectives retrieved the most recent GPS ping of the car's location, which pointed them to a used car lot just north of the city. Detectives were rewarded when they arrived at the dealership and obtained security camera footage of the suspect selling the car—behavior that was suspicious enough to establish a warrant for his arrest. One detective on the case told me, "That was great evidence for us, but I was trying to figure out where he came from or where he went to after, to try to track him down." They called PSS to see if they could get a fix on where he might be laid up.

This was an unusual request for the spy plane program. The actual crime in this case not only occurred at night, when the planes were not allowed to fly, but also on a day when the plane didn't fly at all because of rainy weather. What is more, the GPS ping at the dealership was six days after the day of the murder. PSS was being asked to look at a non–crime scene location on a day that was far removed from the time of the incident. According to the MOU, PSS should have stopped their part of the investigation at this point. On the other hand, police already had a lead on the target, and so one could make the argument that, by his movements, he was pulling new times and locations into the boundaries of what might reasonably be considered "the crime."

Working in the heat of the moment, and not wanting to disappoint investigators, analysts did not stop to consider these boundaries, but immediately went to track the suspect. Pulling up spy plane footage from the day the suspect sold the car, they found that the dealership lay literally yards outside the northern edge of the imagery. If the plane had been flying just the slightest bit farther north in its orbit, it would have captured the car lot and been able to easily sync to the time stamp of the AcuraLink data. No such luck.

PSS remained undeterred, however. Rather than stop the investigation at this point, which they surely should have, they went back to detectives to ask if there were *any other times or locations related to the case* with which they could help. Now they were fishing. Detectives gave PSS the suspect's mother's address and another address of an apartment associated with the suspect, asking if analysts could determine at which location the suspect was staying. "Can you guys watch these two addresses?" a supervisor recalled being asked by detectives.

This request was now clearly outside the standard scenario for a spy plane investigation. It did not even entail a specific temporal data point, let alone a crime scene—just a couple of addresses. Analysts were being asked to watch the home of a person, most likely an older Black woman, not directly involved in committing a crime and not even present at a crime scene, on the off chance that her home might be connected to a suspect's current whereabouts.

The request did not sit right with everyone in the office. As they were putting together the briefing document on the case, Mya, one of the younger analysts, noticed the crime occurred at night and their investigation was of a different set of locations than the reported crime scene. "Do we have authorization to look at this?" she asked McNutt. "Yes," he replied quickly, "it's connected to a murder, we can look."

Not everyone shared McNutt's opinion, though. Jessica, the head analyst supervisor, was unsure. "I thought the requests would fit into these boxes, and they're not," she told me worriedly. Indeed, back in March, before the program launched, I had interviewed Jessica about how analysts would know who they could track and who they couldn't. At that time, she had few concerns because, she said, "We are not planning to do network analysis," referring to an investigatory style that involves mapping the entire social network of a suspect. In her mind, if it wasn't about looking at a crime scene, the company wouldn't be asked to help. Now, here they were being asked to do just that.

Jessica told me she felt uneasy about pushing ahead with the case, but, following as closely to the MOU as they could, she and the other supervisors sent the request up the chain of command to the commissioner to determine if they were allowed to proceed. He signed off on the request, she told me, under the "exigent circumstances" clause. In contrast to what was stated in the MOU, where supplemental requests are meant for something unusual and cataclysmic, the commissioner allowed detectives to use the plane more broadly and begin looking into suspects' secondary networks of association. This is how the program began to sweep more people and places into the scope of surveillance. As I showed in chapter 4, this is standard operating procedure for the BPD. Time and again, they have used experimental surveillance technology, like stingray, that seems targeted at first, but ends up exposing entire neighborhoods to police contact.

With permission from the top, Zach, another supervisor, took the lead on the tracking. He immediately ran into a problem unique to this sort of social network surveillance: because analysts were not tracking from a known crime scene, detectives had given PSS a location but no time at which to start surveillance. How would Zach know when, in the spy plane imagery, to start looking at the suspect's mother's house? Fortunately for Zach, this woman's house lies in an area of West Baltimore clustered with CitiWatch cameras because it sits right outside a public housing complex. One camera panned right by the entrance to the house. Zach reasoned that if he could find a car that matched the suspect's car through CCTV, like finding a needle in a haystack, he could use this to start his investigation in iView and establish a pattern of movement that would suggest the suspect was staying at this location.

Zach proceeded to spend hour after hour rummaging through CCTV footage of the mother's house on the day the suspect sold the car. Remarkably, after half a day of work, he found what he thought was the white Acura of interest departing the mother's address a few hours before the car was sold. Switching to iView, he then tracked the car forward in time as it meandered through the city, stopping at the other location given to PSS by police, before ultimately going to the far northern edge of the imagery, just down the street from the used car lot. Though Zach could not see the car pull into the dealership because that location was yards outside the imagery, the pattern of movement made him certain enough that he, indeed, had the car associated with the homicide scene.

Zach then began pulling up CCTV footage from other days prior to the sale of the car at both the mother's address and the apartment given by police. Going back in time six days, from the day of the car's sale to the day of the murder, Zach was able to establish that the white Acura parked in the same spot, day after day, in front of the second address given to them by police. By establishing this pattern, Zach was able to tell detectives that he thought the suspect was not staying with his mother, but rather at the apartment.

Talking to the lead detective a few weeks later, I was told that they ultimately located and arrested the suspect and would be charging him with murder. I asked if PSS helped in the arrest. Not wanting to sound rude, the detective replied,

> [Zach] was able to find out where he . . . came from just prior [to selling the car] by just going through all those cameras to figure it out, but by the time we got that back, you know, obviously [the suspect] was not at that location anymore. . . . I wouldn't say it—it did gave me peace of mind at the time that we had the right person, but I wouldn't say it was helpful to the investigation.

The detective conceded that Zach's work was mostly in vain because "you know, [the dealership] just happened to be just yards outside [the imagery]." What would have been really great was if the spy plane had just seen the actual crime—the whole point of the technology. Despite working incessantly to find something valuable to contribute to the case, pushing past the privacy boundaries set up in the MOU, PSS ultimately confirmed what police already knew.

The Four-Day Rule

The Hutchinson Street investigation, while painstakingly ineffective, helped open the door to a new use of the spy plane, which became popular with detectives: supplementals. Ultimately, it put the program in legal hot water.

Shortly after this investigation, PSS developed what they came to call the "Four-Day Rule." More a rule of thumb shared between analysts than an official policy hashed out with the BPD, it said that if a request for service did not lie in the plane's coverage area (either by time or place), PSS could still track people *so long as the target lay four days before or four days after the date of the incident.* This "rule" was inspired by one of the most controversial recent Supreme Court decisions around privacy: the *Carpenter* decision. The way constitutional law shaped PSS's behavior in the heat of the moment, then, is crucial to understanding how the program came to break its own rules around privacy.

Performing Privacy

As legal scholar Ari Waldman notes, privacy is a kind of "serious performance" inside tech companies. Though they often publicly claim to "value your privacy," in practice they are duty bound first to profit. Much of what companies produce in terms of actually enacting privacy *law*, then, is about "symbolic compliance"—creating routines inside the organization that make the company appear to address privacy law but without threatening the company's profitability, efficiency, and market position.[8] This is a good description of how PSS approached privacy.

PSS had a particularly difficult time pulling off privacy performances around supplemental cases because these cases launched them headlong into one of the murkiest areas of privacy law: the Fourth Amendment search. This area centers on whether and how the government can use its power to search Americans and their stuff without first getting permission from a judge in the form of a warrant.

Garth, the company's lawyer, said he used a recent Supreme Court decision, known as the *Carpenter* decision, as a guide for navigating supplemental requests. As I discuss more below, the justices in that ruling said that police could not look more than seven days backward in time into a suspect's movements using location data without a warrant. Garth realized they should make sure to avoid that problem, because the plane could theoretically look at any slice of a person's past that was recorded, even if it was a slice that happened long before the person was a suspect. So, they set the limit at four days. Hence, the Four-Day Rule. "I sort of came up with it on the fly," Garth said. "*Carpenter* says you can do seven days, and four is less than seven, so . . . [*laughs*]." As Garth's levity suggests, PSS was concerned with the *How far back in time?* question, but it didn't slow them down. They reacted in the moment to come up with a performance that would symbolically comply with *Carpenter* in a civil suit, should it ever come to that, and pressed forward. Boy, would they come to regret that.

Seemingly protected by a convincing privacy performance, then, PSS had opened themselves to doing investigations on any person, place, or thing tangentially related to a crime that fell inside a span of four days. This was a far cry from the kind of boundedness the BPD had promoted to community members.

Looking at cases that spanned that entire four-day window reveals how supplemental surveillance could sweep an increasingly large number of innocent people into an investigation. In one case, the BPD asked the company to track a vehicle that was connected to a homicide and gave them "a list of locations flagging the suspect vehicle in license plate readers [LPRs]" that spanned nine days. Analysts stuck to the Four-Day Rule and followed up only the LPRs that happened on days right around the homicide, synchronizing the LPR time stamps with spy plane imagery. Analysts noted each address where the car stopped, observed people getting in and out of the car or congregating around it, and other details of everyday life. They ultimately identified "three separate locations that are frequented by the subject." On one day, they used archived CCTV footage to observe "a female . . . getting into the driver's side" of the suspected car and driving to "a mall, a market, and a gas station," before stopping at an address in West Baltimore.

Throughout their write-up of this investigation, analysts use the word *subject* to describe the vehicle's movements. But a car is not a person. It is just a proxy for the person of interest. How could analysts know if the woman driving the car was the same person who drove the car the day before, or, most importantly, at the eventual homicide scene? Maybe she was just borrowing the car to run some errands. And given that the surveillance was two days *before* the crime, there is no way she could have known the car would appear at a *future* crime scene. Did she "deserve" to be watched by the state for this? While the investigation "only" spanned three days, then, even within that short window of time analysts could bring multiple addresses and drivers of the car into the investigation without reasonable suspicion that the person was actually involved. They watched this woman, for example, go shopping, but what if she had been doing something much more private, such as visiting an abortion clinic? Would it have been right for the state to see that private behavior, even though it had nothing to do with the *reason* she was being watched in the first place?

It's important to highlight that this and other supplemental investigations sprawled primarily because police had so much other surveillance data to provide to PSS that did *not* fit within the company's Four-Day Rule, nor restrict itself geographically to the actual crime scene. In a city like Baltimore, there are LPRs, CCTV, residential cameras, cell phones, public bus cameras, fuel pump cameras, and other technologies continually recording and creating databases to which PSS could link. Even though PSS "cared about privacy" and stuck quite strictly to its Four-Day Rule, its investigations could quickly expand well beyond the crime scene to the whole of the city *because it could connect the dots* and rope in locations and people that were degrees removed from the incident.[9]

In one case, PSS was asked to track a car that was involved in a homicide in March (before the program had even started) because in July, according to detectives, the same car was involved in a high-speed chase with police. Multiple suspects crashed the car and fled the vehicle on foot. Unable to locate them fleeing from the crash (which is not a violent crime and is therefore technically not within the scope of the program anyway) in iView, police then gave analysts LPR hits on the car going back *eighteen days*. Chasing up many of these hits in iView, analysts ended up tracking over twenty vehicles that visually resembled the suspect's vehicle but could never positively identify it. In their internal documentation of this investigation, analysts noted that "the cops have the tags of this car running through LPRs for months. This proved fruitless, but please continue to utilize this." With this and other supplemental investigations like it, I got the sense that PSS was engaging in tracking that even they knew was unlikely to yield anything substantial, but they felt unwilling to say no.

PSS's failure to hold the line on privacy policies is explained, in part, by the company's profit-oriented motivation. But it is more subtle than a story of corporate greed. McNutt wanted to stay in the city to continue his personal mission of helping out in the crime fight. "I do not care one iota if I make a dime off of this program," he often said to me. As a for-profit contractor, however, that longevity was dependent on a transactional relationship with the BPD. To stay, they had to please the company's primary customer—police—or risk having their contract lapse. Would pleasing police make PSS a lot of money through a long-term contract? Surely the answer is yes, but that end was one step removed from McNutt's primary motivation of making a long-lasting impact on crime. The profit motive here is not the same as greed, then. It is embedded in the *organizational structure* of the relationship between PSS and the BPD. Even while McNutt and others were trying not to be greedy, it was the structure, not the individuals, that put profit above privacy.[10]

Stepping back and looking at the company's work on supplemental investigations, I see two important sets of questions: First, given that many of the people swept into investigations by supplemental surveillance never knew it, was harm done here? If so, what kind of harm? Is an "invasion of privacy" the right language to describe this harm? Second, given that many of the supplemental investigations did not, in fact, give police a kind of "superpower"—a time machine that can see every intimate detail of someone's past[11]—was harm done here? In other words, if the technology that is supposed to give police an all-seeing eye turns out to be far from total, is there still a problem? In order to address these questions, I need to take you on a detour through the topsy-turvy world of contemporary Fourth Amendment law.

The *Carpenter* Decision

In April 2020, as PSS was still constructing its new operations center, a coalition of racial justice activists and the American Civil Liberties Union (ACLU) sued the Baltimore Police Department over the spy plane program. They argued that the spy plane would give "the BPD a virtual, visual time machine whose grasp no person can escape."[12] This violates the fundamental right to privacy enshrined in the Fourth Amendment, they argued, because it "will capture the whole of an individual's movements and thereby reveal their privacies of life."[13] They sought an injunction from the District Court of Maryland that would put the program on hold.

In May of 2020, they lost that challenge, which allowed the program to carry on. In the meantime, they appealed the case all the way up to the federal Fourth Circuit Court of Appeals. The court met twice (including a second *en banc* hearing) to consider and reconsider the constitutionality of the spy plane.[14] This whole process took over a year, during which time PSS carried out dozens of supplemental investigations using its improvised Four-Day Rule.

Finally, in June of 2021, after the mayor of Baltimore had already decided to discontinue the program, the Fourth Circuit overturned the ruling at the district level, saying that the spy plane was, indeed, as invasive as the plaintiffs originally argued. It was a narrow decision, with eight judges in favor and seven dissenting. Even though the program had ceased gathering new data by that time, the majority judges were adamant that the program should never have even happened.[15] Because the program "opens 'an intimate window' into a person's associations and activities," Chief Judge Gregory wrote in the majority opinion, "it violates the reasonable expectation of privacy individuals have in the whole of their movements" and thus violates the Fourth Amendment.[16] A big target of criticism in the decision was . . . you guessed it . . . supplemental requests.

The Time Machine Problem

Throughout the oral debate over the spy plane, lawyers and judges relied heavily on the Supreme Court's arguments developed in *Carpenter v. United States*. In that case, the justices were asked to determine whether police violated the Fourth Amendment rights of a man named Timothy Carpenter when they repeatedly and persistently accessed information about the location of his cell phone. Employing cell site location information (CSLI), police used Carpenter's phone data to associate his movements with the locations and times of a string of armed robberies. Sound familiar? The case is remarkably similar to spy plane analysis, but it features a database of cell phone records instead of a database of aerial photographs. You can see why Garth, PSS's lawyer, was worried about this decision and tailored the company's Four-Day Rule around it.

The case raised some fascinating questions: What constitutes a "search" by the state when what is being looked at is a digital representation of a person's location? Is a "virtual search" a radically different thing from a physical search? Oftentimes in these cases, it's not like police are looking at one location, as they do in a traditional stakeout; they are looking at dozens of locations repeatedly over a long period of time using digital information that is quick and easy to access.[17] How far back in time are police allowed to look in a "virtual stakeout"? How many times can they take a digital measurement of a person's location before it adds up to government overreach?

Ruling by the narrowest of margins, five in favor and four against, Chief Justice Roberts authored a majority opinion that rocked the world of privacy law. He argued that "individuals have a reasonable expectation of privacy in the whole of their physical movements."[18] In other words, most "reasonable" people (whatever that means) probably think that the government is not able to see all the movements they make over the course of a long stretch of time. Sure, a cop can track me for a short stretch of time; but all day, every day, for days and days—that's just too much.[19] This means that even though police might search for and collect one piece of data that, by itself, seems not that invasive, if they do that dozens of times, exhaustively and persistently, it can add up to an invasive and unconstitutional search.[20]

Moreover, the whole point of the Fourth Amendment, a creature of the eighteenth century, was to prevent the *arbitrariness* of state power.[21] It's not just that the government is searching for and seizing information about your location, it's that they don't have to furnish a reason when they do it and reveal precisely how they will do it. That's how the British used to do things in prerevolutionary America, and that's why the Fourth Amendment exists in the first place. So, the court concluded, if the government wants to gain that kind of arbitrary, exhaustive knowledge of movement, it needs to slow down and first get a warrant.

The natural next question, then, is *How much tracking is too much?* This is what I call the time machine problem. If the state can "go back in time" to look at our movements, how far back should they be allowed to go? How should police, prosecutors, defense attorneys, and judges draw the line in time? Are we talking days here? Hours? And what if the time frame is discontinuous, with some tracking, then a gap for a while, and then some more tracking? Does continuity matter?

Sticking close to the specifics of Timothy Carpenter's case, Supreme Court justices provided almost no guidance on how to draw this line, only noting that police accessed seven days' worth of Carpenter's CSLI data (and only used two of those days in their investigation, actually) and that was too much. So is it too much because of seven or two or some other number of days?[22] Is it the continuity of those days that's the problem, the number of individual hits in the dataset, or what? As I observed in my fieldwork, PSS took these numbers seriously when they constructed their "privacy performance." The specifics of the *Carpenter* case actually structured what the company did with its tracking in the heat of the moment of figuring out how to handle supplemental requests, ultimately resulting in the Four-Day Rule. They limited themselves to four days "because four is less than seven," when they could have picked just about any other number (including seven or two). Talk about the arbitrariness of a government search!

The significance of the *Carpenter* decision is larger than its designation of our patterns of movement as private, though. Throughout Roberts's opinion, he continually highlights that the kind of surveillance carried out on Carpenter was a problem because of the exceptional and unprecedented power of "time travel" given to police by new technologies. Things like GPS tracking and CSLI analysis "give the Government near perfect surveillance and allow it to travel back in time to retrace a person's whereabouts," he writes.[23] Roberts seems awed by technology, here. As legal scholar Paul Ohm notes, "The beating heart of the *Carpenter* majority opinion is its deep and abiding belief in the exceptional nature of the modern technological era."[24] The opinion gives one the impression that new digital devices "transform police into crime-fighting robots outfitted with superhuman powers."[25] Justice Roberts writes with "a palpable, wide-eyed amazement at the speed with which the power and scale of technology has changed" and uses "words and phrases [that] would seem more at home in science fiction than the U.S. Reports," invoking "time travel, space travel, and visits from Martians."[26]

Roberts's opinion in the *Carpenter* case deeply shaped the way Judge Gregory, of the Fourth Circuit Court of Appeals, looked at the spy plane. The crux of Gregory's argument was to stop looking at the plane's imagery as aerial photographs, as BPD lawyers had done, and start looking at it as "data." The spy plane creates an archive that resembles a location database, like CSLI or GPS, more than it resembles other aerial surveillance photography and video. This shift in reasoning from the literal visual image to the abstract database was crucial to roping the spy plane into the approach to privacy developed in *Carpenter*. Rather than raise concern about how clear or fuzzy the imagery is and whether it can look inside your house, as BPD lawyers had emphasized, the court focused on how the technology purports to catalogue movement precisely and persistently. Looking at it that way, Judge Gregory argued, the spy plane is overly invasive. He noted that even though the spy plane can't see what you look like or peer into your house, it can still be used to piece together long stretches of daily movement that open "an intimate window into a person's life."[27] Concluding in an almost poetic mode that grants WAMI an almost magical invasiveness, Gregory argued that the spy plane is akin to giving the police a "21st century general search"—a digital version of the kind of power the Red Coats wielded over US colonists, where they could roll up unannounced, knock down the door, and take whatever they wanted.[28] Yikes.

The Harm of Experimentation

For me, watching the privacy debate about the spy plane unfold was maddening because it constantly played into dystopian "doomer" hype. During the Fourth Circuit's oral arguments, I frequently found myself yelling "It can't do that!" or even "Oh, police *wish* it could do that!" On the one hand, I saw many worrying things unfold as the program cracked the MOU and oozed out of its established boundaries for privacy. On the other hand, the spy plane is not a "virtual, visual time machine" that can miraculously capture the "whole of an individual's movements."[29] It is similar to "attaching a GPS to every person" or tracking a cell phone, but not quite: you lose the signal when it's cloudy or when the person goes too far in a particular direction. There are gaps in the dataset between days that create discontinuity. Moreover, because the program relies so heavily on other surveillance sensors, a track might go through parts of a city with no other cameras or data that *could* actually reveal "the intimacies of life" about a target. Even when it does, there are a thousand difficulties in connecting all these sensors in a way that could actually create a perfect "digital dossier" on an individual. And, to put it bluntly, the spy plane imagery is terrible—it is fuzzy, jumpy, and difficult to interpret. It is so much more imperfect than the doomers claim. It is not an all-seeing Big Brother. It is more like Henry behind the frosted glass door. That's not any less creepy, but it is certainly less mechanically powerful.

If critics are arguing that the harm done by the spy plane was that it created a relatively seamless location database that could reveal the intimacies of life of every Baltimorean, then that's just empirically inaccurate. The critique is actually a form of hype.[30] That is not to say that the program didn't overstep its boundaries in problematic ways. In debates over privacy and police technology, however, rarely is there a middle ground.[31] The technology is either dystopia or utopia. The problem with this is that both sides end up reproducing the sentiment that technology is magic—that it can create a perfect record on every person and serve it up to government eyes with the snap of a finger.

This hype even shapes some of the most compelling advancements in thinking about privacy. To take one important example, Julie Cohen's groundbreaking and deservedly celebrated intervention in privacy theory argues that we live in "an age characterized by . . . increasingly precise efforts to monitor and predict individual behavior with comparable seamlessness and granularity."[32] The information systems that surround us have "the ability to identify individuals persistently and accurately." They are closing in so tightly, she argues, that we lack "the breathing room that critical subjectivity requires." She calls for a new set of design principles for data systems that intentionally include more gaps, discontinuities, and even inefficiencies, in order to allow the expressive subject room to play outside the all-pervasive gaze of corporations and the state. While I wholeheartedly agree with these political goals, the characterization of what technologies can *actually* do sounds overly dystopian (or utopian, if you have different political leanings). There are *efforts* by tech companies and police to create such precision and granularity, for sure, but it doesn't seem to me that they actually translate on the ground.

This mismatch between what companies and police are *trying* to do, what they *say* they can do, and what they *actually* do is a problem for debates over privacy and surveillance. In the case of the spy plane, one could argue that PSS actually followed Cohen's principle of engineering gaps, discontinuities, and inefficiencies in the technology. They said they would do only one pixel per person, even though they could do more; only look at known crime scenes, even though they could look anywhere. They did this not to protect privacy, however, but in order to make the technology seem less creepy. In fact, it was sometimes *because of* these gaps, discontinuities, and inefficiencies that the technology ended up invading the privacy of people who didn't "deserve" it. The suspect's mother in the Hutchinson Street case, the woman driving a suspect's car who was tracked while shopping, the dozens of people driving cars that resembled one from a list of LPR hits—the company seems to have known that much of this tracking was aimless and fruitless, but did it anyway out of a panicked desire to make the technology seem more useful than it was. The reason these people came under the gaze of the state, then, is not because of some superpower that allows the state to see into our souls, but because of the weak privacy *performances* the company engaged in while experimenting with a glitchy and limited prototype.

How, then, should we talk about the sources of privacy harm produced by the spy plane? If we use the doomer frame and think of the spy plane as some sort of existential threat to humanity, we would argue that the harms to privacy are because of its time machine–like capacity to let the government see into *everyone's* personal lives to a shocking degree. The reason the spy plane would be harmful is because, as the ACLU argued, it creates something as creepy as a virtual cop following you every time you step outside your door. As I hope is now clear, that's criti-hype.

I think the spy plane's privacy harms are better understood through the lens of experimentation. One of the payoffs of thinking this way is to highlight how surveillance systems are actually used, right now, in the present.[33] More often than not, that is in a prototype capacity, aimed at a small subset of race-class subjugated communities that have been offered up, often without consent, as test subjects. The risks that might come from this experimentation—false positives, unnecessary tracking, and so on—are not an "existential threat" to a future humanity, but a present threat to people in Black and Brown neighborhoods.

The legal balancing we should do, then, puts inequality squarely in the frame. We don't have to debate whether the technology is *actually* as powerful as it is being made out to be. All we need to know is whether the experiment is designed justly or unjustly. Sure, the technology *might* someday get "better" and truly be able to "watch us all," and that would be really bad. But to get there, think of all the lives that would have to be experimented upon to figure it out. To be clear, I think we absolutely should create legal arguments to head off technologies that are an existential threat to humanity. What I object to is that, oftentimes, this is *all* we're talking about when it comes to privacy. Shouldn't concerns about harms that are happening *right now* be at least as urgent as some dystopian future full of supercops?

Here is the main point, then: the spy plane's privacy harms *initially*, before it ever gets as powerful as a time machine, issue not from the problem of invasiveness but from the problem of experimentation. I don't think the spy plane creates a database that can reveal the intimacies of every single American, or even Baltimorean. I think it is an experiment on Black lives—lives that are already destabilized by poverty and racism. As I explore next, this experimental program then hands its evidence over to a court system, already ill-equipped to deliver real justice or give closure to grieving families. This creates real concrete harms, not in some potential future, but right now.

7 Mechanical Witness

Sergeant Charles and his detectives pile into the claustrophobically small briefing room inside the spy plane operations center with all of Team 3, as well as McNutt and several supervisors. We're packed like sardines. The suspect in the Carroll Avenue homicide, which occurred a few days earlier, had just been apprehended with a warrant for murder after a brief foot chase. As described in chapter 2, this arrest would have been impossible without the spy plane. The case was shaping up to be a massive success for the program. But now the real work would begin. Could this arrest be turned into an actual conviction in a court of law? Would the evidence from the spy plane hold up in front of lawyers, judges, and jurors? As the program's boosters hoped, could PSS "send a message" through the court system that you can't get away with murder in Baltimore when the spy plane is up?

Charles has come to PSS to triple-check their investigation because he is worried about how prosecutors will react when they put the case forward. It became clear early in the pilot program that the state's attorney, Marilyn Mosby, did not like the spy plane. I talked to a former Baltimore prosecutor who had firsthand knowledge of this. He told me Mosby had sent a clear message to all her attorneys that WAMI was a risky and untested technology. She warned the office that if prosecutors wanted to pursue a case based on spy plane evidence, "It had better be open-and-shut."[1] Anything shy of this would be setting prosecutors up for failure in court.

Charles asks McNutt to walk him through the whole Carroll Avenue investigation again, just to make sure it will pass muster with the prosecutor. McNutt goes to the front of the briefing room where there is a big projector screen. Looking a little nervous, he puts the evidence packet up—basically a PowerPoint deck with still frames from the footage and analysts' interpretations. This is one of the main services PSS advertises to its law enforcement clients. They don't just provide the technology and walk away. PSS comes with a team of experts who can, like any other forensic experts, provide interpretations of a scientific tool that can help prosecutors build ironclad cases. McNutt launches into his presentation, showcasing PSS's bespoke analysis, but Sergeant Charles quickly cuts him short with, "Can you just pull up the video?" Charles seems impatient and uninterested in McNutt's slides. McNutt pulls up iView, loads the Carroll Avenue footage, and zooms down into the time-space of the crime scene. He begins to walk Charles through the tracking, step by step, showing how analysts were able to keep a nearly unbroken visual fix on the suspect from the time he woke up that morning, to the murder itself, to when he ditched the stolen vehicle used in the crime. After a long silence, Charles looks satisfied. "This is great," he says approvingly.

It is crucial, he says, that they be able to convince the prosecutor that no other person could have gotten into or out of the car—that there is just one suspect, one vehicle. So, Charles tells McNutt to run back through the time frame just before and after the murder one more time.

There is a hiccup. About fifteen minutes prior to the murder, we see the suspect's vehicle heading toward the murder scene, when the car pulls over to the side of the road and stops for about five seconds. "Wait, what's that?" Charles points. This stop had not been discussed before. He sits forward in his chair. "Is anyone else getting in the car there?" he asks. There is a pregnant silence as McNutt zooms in closer and runs the imagery backward and forward several times and says, "Well, there might be someone walking on the sidewalk right there, but I can't tell for sur—." "Don't say any of that [to the prosecutor]," Charles aggressively cuts in, stepping over McNutt's last syllable, continuing,

> Just show that he pauses and moves on. Make it seem like you are testifying in court. Let her ask the question. If she doesn't ask [about this stop], just let the video play. If she says, "What are they doing here?," don't give your opinion. Just say what you see.

This portion of the footage could spook the prosecutor, Charles worries, because there is the smallest indication of a second person being involved. The whole case hinges on a simple story—one gun, one suspect, one car. Anything suggesting more will tank the case. The work of detectives to generate a simple "prosecutable case" and the work of the spy plane analysts to generate mountains of visual evidence are in deep tension in this moment. This tension—between what police *want* the spy plane to see and what it *actually* sees—led Charles to make what is clearly an ethically dubious decision: instructing analysts to change their analysis to suit the police narrative.

Part of the public appeal of a technology like WAMI is that it can create objective evidence by functioning as a kind of "mechanical witness"—an independent observer of a crime scene that is free of human biases. As I have been at pains to show, however, the spy plane cameras simply record pixels; human analysts do the witnessing. There is nothing purely mechanical about it. Watching how detectives steered McNutt and PSS analysts during the Carroll Avenue investigation made me realize that these interpretations can be influenced by the motivations of police and prosecutors. Moreover, this bias toward the perspectives of police is baked into the structural location in the criminal legal system that PSS occupies. PSS was working so closely with detectives and prosecutors, seeing them as the primary client they needed to please, that they were incentivized to side with police interpretations. In this instance, when the choice between full documentation and a simpler narrative that would perform better in court was presented, PSS felt pressured to cave to what was expedient for police and prosecutors. This feels like a slippery slope. If analysts are willing to bend their interpretation of the imagery on this detail to get in line, what else would they be willing to compromise on?

In the end, the concerns about the quality, reliability, and usefulness of spy plane evidence in court were all for naught. When the Fourth Circuit Court of Appeals handed down its decision in June 2021 ruling the program unconstitutional, the State's Attorney's Office (SAO) threw out all the cases that were touched by the spy plane. Prosecutors knew that whatever evidence the spy plane pointed them to was tainted. On the surface, then, that's the end of the story: the spy plane is unconstitutional, and that's why all the cases failed in the prosecution phase.

There is a deeper story here, though. For three years, I followed the evidence from the Carroll Avenue investigation and several others as it meandered through the maddeningly slow Baltimore court system. I listened to hearings, watched suspects waive their right to a speedy trial through delay after delay, and interviewed attorneys to find out what it was like to deal with this evidence. In this chapter, I describe two failed homicide prosecutions, showing how the spy plane program upended people's lives, both those of the suspects and the victims' families. These cases provide concrete evidence of the harms caused by experimenting on Black lives with the spy plane. But this long process revealed an even deeper set of questions. Watching how these cases failed in court made me wonder what would have happened had they gone forward? What if these cases *had* been prosecuted further and even gone to trial? Answering this hypothetical, I speculate that, as evidence, WAMI imagery is so difficult to use in a US courtroom that it may be more trouble than it's worth. The difficulties of its use may create conditions ripe for unethical and even corrupt behavior to grow.

Fruit of the Poisonous Tree

After the Fourth Circuit ruled on the spy plane, the SAO seems to have thrown up its hands. Using a process called *nolle prosequi*, prosecutors voluntarily chose not to move forward with these cases. Just as had happened a few years before with stingray technology, a new and untested surveillance tool had created a mess in the courts. It's mere presence in the casework was a problem because the program's unconstitutionality would be such a glaring target of criticism by any competent defense attorney. Even if the most compelling evidence against a defendant didn't come directly from the plane, the overall investigation still involved unlawful searches. The whole thing was ruined.

The Carroll Avenue case, described in chapter 2—perhaps the most successful during the investigatory phase—was thrown out in June 2022 (almost two years after the murder). The McHenry Street case, described in chapter 3, which featured two deeply troubling false positives, was thrown out in November 2022 (two years and one month after the murder).

My conversations with the prosecutors and defense attorneys who worked these cases made it clear that it was the unconstitutionality of the technology alone that quashed them. Attorneys never even got to the point of deeply looking at the actual content of the imagery to see if it made any sense or would convince a jury. Xavian Miller, a veteran attorney with the Office of the Public Defender

(OPD), was assigned to the McHenry Street case in February 2021, already a year after the incident, by lottery. "We literally spin a wheel," he laughed, "and when it's your turn to come up, it's your turn." It wasn't until May of that year that he would receive the evidence packet from the SAO, through the traditional discovery process. "I was going through the detective's notes and I saw that there was this AIR research team that had been involved pretty early on, like, day of the incident," he recalls. He and all the other attorneys in the OPD had been notified by the forensics division to "be on the lookout for these cases." He reached out to the forensics expert for advice. As discussed in chapter 3, the prosecution's case in the McHenry Street murder hinged on PSS tracking that showed the suspect's blue Cutlass stopping at a gas station hours before, and a mile away from, the incident. When Detective Karim, the lead investigator, took video from the gas station into the interrogation, the suspect unwittingly admitted that the pixelated dot in the corresponding aerial footage was him.

The forensics expert at the OPD advised Miller to try to have the gas station video thrown out, because it had been acquired through unconstitutional means, knowing full well that this would completely unravel the prosecution's case. In January 2022 (over a year since the murder), Miller filed a motion to "suppress evidence recovered as a result of the unlawful use of images/video from the AIR program." Miller didn't have to do much in his motion to back up his argument. He literally cut and pasted from the Fourth Circuit's majority opinion, which had just been announced a few months prior. "Because the AIR program enables police to deduce from the whole of individuals' movements," Miller's motion read, "accessing its data is a search, and its warrantless operation violates the Fourth Amendment." Miller then applied this principle to the gas station footage: "Because officers used the data from the AIR program to secure footage at a gas station they would not have otherwise obtained from an independent source, that evidence is also the fruit of the poisonous tree." The identification of the suspect from that footage, then, "is also the fruit of the same poisonous tree and ought to be suppressed." If the spy plane itself is legally poisonous, then it doesn't matter what the evidence pointed to by the plane shows; it was obtained from a toxic source. The whole case should be thrown out.

About a month later, prosecutors tried to save their case, filing a response to Miller's motion. They argued that, sure, the spy plane was ruled unconstitutional in June 2021, but the McHenry Street investigation took place long before then. How can you say that the investigation was unlawful when police were using technology that, in the fall of 2020, wasn't yet considered unconstitutional by any court? In fact, the first time the program was looked over by the Fourth Circuit, they deemed it lawful. It would require a second *en banc* hearing before that earlier decision was overturned. "In October of 2020, Officers reasonably could have thought that the AIR program was constitutional," prosecutors wrote, "and were thereby acting in good faith when utilizing this data." Melissa Rogers, the lead prosecutor on the McHenry Street case, admitted to me that, inside the prosecutor's office, they were taken aback by the Fourth Circuit's decision. "We thought the whole concept [of the program] was great," she recalls. Everyone in the office thought "there are cameras on light poles all over the city and on traffic lights." The spy plane is "no more invasive" than this, she said, adding, "It came by surprise to us" that the program was struck down.

Rogers notes that, though her team filed this response to Miller's motion to suppress, she ultimately didn't argue the point further in an official hearing because of a development in another important spy plane case—Carroll Avenue. Rogers learned that the same judge on her case had already struck down spy plane evidence in the Carroll case earlier that year.[2] Seeing the writing on the wall, the SAO decided to drop all cases that had even a hint of connection to the spy plane.[3]

The fact that *these* cases—broad daylight murders in one of the murder capitals of the world—failed after years of delays was both embarrassing for the Baltimore court system and traumatic for victims' families and friends, as well as the accused. It's important to understand the impact on both groups. The defendants sat in jail upward of two years while attorneys figured out how to deal with the complications from the evidence. Between long stretches of incarceration, defendants appeared at sporadic hearings over remote video chat. They were asked to waive their right to a speedy trial because council needed more time to figure out what to do. Cases were postponed multiple times after having been scheduled for jury trials. One defendant became ill from COVID-19 during his time awaiting trial. After the Fourth Circuit decision in June, he filed three motions, handwritten from jail, repeatedly pleading with the judge to dismiss his case. "Whereas aerial surveillance is used after it was banned and deemed to be illegal," he wrote the judge using large, carefully formed letters on lined notebook paper, "thereto the state loses its probable cause to charge and hold me in all counts of the indictment." He would rewrite and refile this same motion two more times, spanning a period of about four months. It was not until June 2022, a full year after the program was deemed unconstitutional, that he was released. He spent nearly two years in pretrial detention.[4]

Not everyone would see the extended time defendants sat in jail as a problem. Many in law enforcement, for example, might argue that it is actually a good thing that these cases took so long. At least the suspects, many of whom had long criminal histories, were "off the streets," preventing them from further threatening the victims' families or committing other crimes.

I disagree with that position. What I saw suggests that these delays were not only bad for victims' families, but their entire *neighborhoods*. According to Detective Karim, the lead investigator on the McHenry Street case, the family of the victim was given high hopes when an arrest was made.[5] They were certain about who did it and had immediately told Karim the man's identity. They were still very skeptical, however, that police would actually find and arrest the right person. This is not surprising, given that many big city police agencies in the United States, including in Baltimore, close homicide cases less than half the time.[6] Karim recalls that when he went to the family's house to inform them of the arrest, they were in disbelief. "They were extremely excited that we closed it," he said. The victim's sister, who answered the door when he came, told him, "No way you got the right guy." To prove it, she asked that he present a mugshot of the person they arrested. She was floored when he showed a photo of the right person. "No way. You guys actually *did* get him," Karim recalls her saying.

Two years later, Karim had to return to the family to tell them the case would be thrown out. Because the family did not respond to his calls, he went in person to deliver the bad news. Expecting the worst, he brought along another officer, "just in case something popped off." Once at the house, he called Rogers, the prosecutor, who informed the family over speakerphone that the case had been dropped. She gave the legal justification about privacy from the Fourth Circuit decision. "We told them that the evidence we had was not gonna hold up in court," Karim recalls, "and without that we didn't have enough to put on the case." Rogers said she gave the family the full explanation of the Fourth Amendment issues. "We told them everything," she recalls.

Karim describes the family's reaction as "extremely upset." "There was a long back and forth and they were asking a lot of questions and there was a lot yelling," he said. The son of the victim, who was "young and troubled and has some mental health problems," was "flipping out," saying that he was going to go out and find the suspect and kill him himself. Karim didn't take this threat seriously, he told me, but he clearly worries about future violence stemming from this case's failure. He pointed out that he was forced to return the very recognizable dark blue Cutlass to the suspect after his release. Because the suspect lives just a few blocks away from the victim's family, Karim thinks there is every chance that they will see the suspect or his car on a regular basis. The likelihood of retributive violence, he explains, is extremely high.

The notion that the delays to the spy plane cases may have prevented violence by "keeping bad guys off the street" for a period of time, then, is doubtful. The program's constitutionality problems delayed things for years. Long bouts of incarceration during pretrial detention only seem to destabilize the lives of the accused more and can lead to worse outcomes.[7] Victims and their families had their cases dragged out and their hopes falsely stoked, and were ultimately led to distrust the criminal legal process even more when the cases fell apart at the eleventh hour. I can see how this kind of emotional turmoil would make it seem completely reasonable to want to take the law into one's own hands. Police and attorneys wasted a lot of time on these complex cases, at a time when they are already stretched thin, further exacerbating a system that fails to deter violence through swift justice. Nobody wins here. Just from these few cases, then, you can see that many people's lives were upended because of an experimental prototype that was unviable in court. And that burden fell heavily not just on individuals but on entire neighborhoods in Black Baltimore. This is what I mean about "concrete harms in the present."

I cannot stress enough that this outcome should have come as no surprise to prosecutors. As detailed in chapter 4, the exact same thing happened just a few years prior with stingray, a similar surveillance technology. Everyone in the criminal legal system could have and should have seen this coming and prepared for it as a highly likely outcome. Instead, the BPD's legal team constructed the spy plane as "just another camera" that is "no more invasive" than what was already in the city.

The lesson we might take away from this series of events is this: the BPD should have taken a bit longer to vet the constitutionality of the spy plane before conducting the experiment. At the very least, they should have only used the plane with a warrant, just in case a civil suit popped up. The problem with the experiment was that it was too rushed for a court system that is so broken and unsophisticated; the technology itself, though, was not really at fault.

I want us to look a little deeper, though. Even after I learned that these cases were thrown out, one thing continued to nag me. I couldn't get over the fact that, at every stage of the criminal legal process, people consistently found spy plane evidence frustrating, complex, and difficult to deal with. As I saw in the Carroll Avenue case, this complexity then led them to make choices that were ethically dubious. If the premise of the spy plane is that it would make the investigatory process that much fairer by operating as an unbiased, objective, independent witness, that's certainly not the sense I got. So, I think it's important to consider the hypothetical: What if these cases *had* gone to trial? How would things have played out?

Mechanical Witness

Moving picture and video evidence have been around the US court system since the 1910s.[8] Looking at this history, one fundamental question has never been fully answered by the courts: are moving pictures to be treated as "just a story" or as an "objective fact"? When a human witness gets up on the stand and tells "the truth, the whole truth, and nothing but the truth," we know what to think about that. This witness is using narrative to tell their *version* of the truth. They can always be cross-examined, and a good attorney can show why the narrative presented is not the real, capital *T*, Truth. What about a mechanical witness? What should we do with moving picture evidence—film, video, or WAMI's even stranger flip-book-like footage? Is a mechanically created image of crime a different sort of thing? Rather than a story, is it more like "raw data"—a passive, neutral record of events?

Because spy plane imagery has never been presented in court, it's difficult to know how it would be seen by a judge or jury. Still, history can give us some hints. One of the watershed moments that sheds light on this question is the Rodney King trial. The brutal beating of King on March 3, 1991, by LAPD officers was caught on handheld video by George Holliday, a bystander watching the arrest from his apartment terrace. Lasting nine minutes, the footage shows four officers beating King on the ground using batons, fists, and legs. Disseminated through television news, the Holliday tape is often considered the first "viral video" of police brutality.

The video was brought by the prosecution into the first jury trial of the officers as *the* key piece of evidence. "What more could you ask for?" the chief prosecutor told reporters about the tape as jury deliberations began. "You have the videotape that shows objectively, without bias, impartially, what happened that night. . . . It can't be rebutted."[9] Just a few days later, however, after spending hours looking at the video almost frame by frame, the jury returned a not guilty verdict for all four officers. Black communities in LA rose up in shock and horror.

In his penetrating analysis of this trial, anthropologist Charles Goodwin showed just how easy it was for prosecutors and defense attorneys to make this video say completely different things. "Opposing sides in the case used the murky pixels of the same television image," Goodwin observed, "to display to the jury incommensurate events: a brutal, savage beating of a man lying helpless on the ground versus careful police response to a dangerous 'PCP-crazed giant'."[10] The officers' legal team, then, simply put on a more convincing display, and the jury adopted that interpretation of the video. How did they do it? They used narrative techniques, Goodwin shows, such as highlighting, coding schemes, and graphic representations to focus the jury's attention on the details of each pixelated image, twisting the meanings of the video toward a conclusion favorable to the defense: police were justifiably terrified of King and did what police officers are trained to do with violent people.

Goodwin argues that the key to the officers' legal strategy was the use of expert testimony to analyze the video, frame by frame, and tell jurors *how* to see its meanings. Through this narrative performance, analysts argued that the video didn't show King's body getting beaten (which it clearly does), but rather King's body making "aggressive" movements "associated" with the behavior of someone on PCP, which, according to "standard police procedure," requires the "escalation" of the "use of force." "Allowing expert testimony," Goodwin notes, "had the effect of filtering the events visible on the tape through a police coding scheme."[11] In the end, it wasn't *what* jurors saw in the video that mattered, so much as *how* they saw it within the powerful schematic framing developed by the officers' legal team.

The King trial, like much of the history of visual evidence in court, suggests that the distinction between subjective narrative evidence and objective "raw data" is false.[12] The utopian dream for visual technologies like the spy plane is that they will give criminal legal actors access to so much more self-evident information about what really happened. Recall that McNutt would often say his plane can capture the "ground truth" of a crime scene. All lawyers would have to do, then, is hit play and let a jury see the Truth.

Looking at the history of moving picture imagery in the courts, however, there is no good reason to think this would be true. The flip-book-like footage could be easily chopped, coded, and reframed, just like the Rodney King video. Even imagery that seems to have far less ambiguity, such as a bystander video or officer body cam footage, can be twisted to match different arguments.[13] Visual imagery, no matter how high the resolution or how supreme the point of view, is always a portrayal from a particular perspective made in a particular context.[14] Clever investigators and attorneys can leverage their positions of power to make the imagery conform to their goals.[15] As photographer Richard Avedon famously quipped, "All photographs are accurate. None of them is the truth."[16]

We can now return to my hypothetical with fresh eyes: Had the spy plane cases gone to trial, how might police and attorneys have worked to chop, recode, and twist the meanings of the footage to fit their agendas? What techniques might they have used? What are the implications for the criminal legal system if a supposedly mechanical witness turns out to be nothing more than another piece of biased narrative?

Brittle Objectivity

Prior to the 2020 experiment, spy plane evidence had never been used in a jury trial. Long before spy plane footage made its way into the prosecution phase, then, detectives seemed unsure of it as evidence. One of the biggest problems they saw was that the plane provided *too much* information on a crime scene. Already overburdened with the sheer number of violent crimes they were responsible for due to understaffing, detectives frequently lamented that the briefing documents from the company were too big and complex. McNutt's team was often instructed to "track anything that moves" around a crime scene. As a result, briefings could sometimes include dozens of tracked vehicles and individuals, each carefully color coded and labeled by number. In a focus group conducted by the RAND Corporation, one detective noted, "All this 'Vehicle 3, 12, 14, 19,' all just seems so overwhelming." "If you have a situation where PSS is tracking information for thirty people," noted another detective, "it's not as helpful. . . . [T]he PSS packet can get really overwhelming with all these variables." The evidence packets were often a mess of colored lines and labels, which even an unrushed detective would have trouble making sense of.

It wasn't just that there was too much information for detectives, though, it was also how this amount of data could lead to vulnerabilities if detectives ever took the stand to testify. One voice inside the company first sounded an alarm about this to McNutt. Shaundria was the lone Black woman and Baltimore resident hired by PSS at a supervisory level. She came with impressive insider experience. She had worked as a beat cop, detective, and internal affairs investigator in the BPD for over a decade. She even worked for the SAO *and* at a defense attorney's office for a few years.

One of Shaundria's first observations about the spy plane, which she raised during her initial day of training, was that the way it swept so many people into an investigation could actually be a problem for detectives because of how it would look in court. During one particularly tense meeting with McNutt, Shaundria presented a concrete example of this problem. In one murder investigation, PSS had identified what they thought might be a witness sitting in a large, boxy ice cream truck parked half a block away, pointing in the direction of the crime scene. Analysts determined that the truck starts up and drives away a few minutes after the moment of the shot and pulls right around the body in an awkwardly wide turn. For McNutt, this was an incredibly important detail. PSS had found a potential witness.

To Shaundria, this evidence was a liability because of what she knew about Baltimore jury trials. Not only would it be nearly impossible for the detective to track down this specific ice cream truck driver, but chances are the person would refuse to talk anyway for fear of being labeled a "snitch." Giving this information to a detective was like sending him on a fool's errand.

But—and this is the crucial part—now that the detective had been told about the ice cream truck by PSS, *he was obligated to check it out.* If he doesn't, Shaundria explained, and this case goes to trial, a defense attorney could easily use the detective's lack of follow-up as evidence of his general incompetence. "Why didn't you find and question that guy in the ice cream truck?" Shaundria said to McNutt, pretending to be a cross-examining defense attorney. "What else did you ignore in this case?" All a competent attorney would have to do is pick the smallest inconsistency or lack of thoroughness in one of the dozens of details given by the spy plane and use that to sow doubt in the jury. That could crash the entire thing. "A defense attorney is going to have a field day with that," she said to McNutt, "to show how messy the operation was. And then you put that in front of a *Black* jury, who already doesn't trust the police, and that's the end of it." She was taken aback that McNutt and others in the company had not noticed this, and even more alarmed when she learned that McNutt's team had *never even sat through a murder trial* in Baltimore to think through how the technology would fit in.[17]

When the problem of too much data arose, I immediately began to wonder just how easy it would be for a defense attorney to exploit other "messy" aspects of the spy plane footage. Given its graininess and halting, jumpy movement, wouldn't it be incredibly simple on cross-examination to contest analysts' claims about what they saw in the footage? All it would take is one small mistake or uncertain interpretation to sow doubt about the program. For all the talk about the objectivity of the spy plane, the imagery seemed brittle.

Brittleness and Corruption

I argue that there is a worrisome link between the brittleness of the footage and corrupt and unethical behavior. The McHenry Street case, for example, featured numerous tracking mistakes resulting from the difficulty of the footage, such as shadowing and gaps in the images. As described in chapter 3, these mistakes initially went undetected, leading to two worrying false positives that sent Detective

Karim to the wrong address. While analysts and detectives brushed this messiness off as par for the course, and ultimately thought of the spy plane as crucial to closing the case, I wondered how a defense attorney would see it.

When I talked to Xavian Miller, the defense attorney on the case, about the mistaken tracking, he was surprised and concerned. He didn't even know the mistakes had happened. "That's news to me," he said.

> That would have definitely been helpful [to know]. It tells me that the technology's not that reliable. . . . It goes to the accuracy—whether the evidence can be trusted. You know, we run into similar problems with the facial recognition, you know, false positives. . . . Had I known that, that would have definitely been a part of my suppression motion.

Miller couldn't include the problem of false positives in his consideration, though, because it never appeared in the records disclosed to him. He thinks that, had the case progressed to trial, he would have eventually asked to see the raw tracking footage (and PSS would have surely obliged to help). There is a chance, then, that he would have eventually taken note of the accuracy problems in the investigation.

Here, then, is where the brittleness of the spy plane imagery becomes fertile soil for the growth of unethical and corrupt behavior. Because detectives find themselves having to do a very difficult job with technology that is untested and frustrating to use, the pressure mounts to do everything possible to suppress these imperfections to prevent the entire case from failing. Detective Karim, who headed the McHenry Street investigation noted that, even though he was aware of the mistaken tracking and agreed that this information *should* be conveyed to the defense, said, "By the time we would have gone to trial, I probably would have forgotten about it." After years of postponements and challenges, little details like this, especially when they are not good for the prosecution's case, could have been easily forgotten (or even conveniently "forgotten").

Karim emphasized that even if Miller's team had seen the raw tracking footage, "they wouldn't have known what was in it. We really trusted the analysts to tell us what's in it." It's unlikely that nonexperts, who cannot really tell what's going on in the footage with their own eyes, would have even known to ask about any mistaken track. The questions, then, are these: Would the PSS analysts who did the messy McHenry Street investigation have voluntarily admitted that they got the tracking wrong multiple times? Would McNutt, a businessman testifying to the accuracy and reliability of his own invention, have fessed up to this? We will never know. What it suggests, though, is that the brittleness of experimental data can lend itself to unethical and corrupt behavior, not necessarily because investigators are ill-intentioned, but because they become desperate to make the technology "work as advertised" in their case.

The Carroll Avenue case, perhaps the most compelling illustration of the spy plane "working" flawlessly, is a perfect example of how brittle objectivity and corruption knit together. When Sergeant Charles saw that there was one unaccounted for five-second stop, and even more frustratingly, McNutt couldn't tell if it showed evidence of another person in the vicinity of the suspect's car, Charles made a split-second decision to suppress that detail to the prosecutor.

Seeing this, I began to wonder just how difficult it was to actually see this detail in the controversial footage. After the meeting with Charles, I decided to take a look for myself at the five-second pause he had instructed analysts to "not mention." I sat down at an unused analyst station, pulled out my key card, slid it into the terminal, and quickly navigated to the spot in question. From my field notes:

On the sidewalk you can make out just the faintest little glimmer of movement just behind the car heading in a Westerly direction. I go through these frames again and again and again, and it just isn't clear enough. It could be a person walking on the sidewalk, but it could also just be some flickering of pixels from the shadows of trees that line the street right there. You could make the case that this is just artifactual movement of the imagery in that area and not the movement of a person. . . . But, you could also make the case that this is a person walking away from the car on the sidewalk.

The imagery is frustratingly difficult to decipher at a particularly crucial point. There might be a person, but there might not be. The case seems rock solid otherwise, but that small imperfection makes the whole thing feel like it's going to fall apart. I can imagine, if I was a detective who only closed a case five out of ten times, I might downplay this small detail out of sheer desperation.

What strikes me as even more worrisome, though, is that even analysts did not note this suspicious stop in the final briefing document. It would take a careful examination of the raw footage, with trained eyes, to notice this absence. I cannot know for sure, and no one would confess to this when I asked, but my hunch is that the stop was never noted by analysts because of pressure from Sergeant Charles to make this inconvenience in the footage go away. Analysts were essentially warned that something as small as this, a few unaccounted for pixels, might ruin the prosecution's entire case because it didn't conform to their theory. This could sow doubt in the jury and ruin the argument. When the pressure gets this high and the experimental evidence is this fragile, people start to make choices that might seem innocuous, but ultimately are unethical or even corrupt.

Given that the spy plane footage is so brittle, my sense is that, had it been presented to a jury, the only way to make the footage truly convincing would be to suppress the data's imperfections. Much like the lawyers for the officers who beat Rodney King, I suspect that any trial would involve chopping and reframing to make the imagery fit a particular narrative. I can imagine that the temptation to hide imagery from prosecutors, as happened in 2016 when PSS investigated the GTTF officers, would be around every corner. The brittle objectivity of the imagery, the frustrating complexity of the system, and the dozens of people contributing to make it seem simple and effective are conditions ripe for corruption.

Parallel Construction

Even though it would be quite tricky to present spy plane evidence in court because of its brittle objectivity, investigators and attorneys would still have another option: neglect to mention that the plane was used, while still relying on the evidence to which the plane pointed. This is a surprisingly common tactic in high-tech investigations.[18] It is known as "parallel construction." Investigators sometimes create two sets of evidence: one for the initial investigation involving practices that are inadmissible in court, but lead to valuable evidence; and another for the courtroom, where the unlawful practices have been removed, leaving only the valuable evidence.[19] Though it might seem hard to believe, investigators can legally lie about and hide that they used warrantless invasive surveillance, if it led them to other admissible evidence that helps their case. On the stand, they can simply say, for example, "We obtained CCTV footage of the suspect's vehicle fleeing the scene" and leave out the part about how they found that with, say, a spy plane. As Nancy Gertner, a former federal judge and law professor, notes, in practice parallel construction is just "a fancy word for phonying up the course of an investigation."[20] Though there is no hard data on it, at least one report suggests this practice has become extremely common in US courts.[21]

There is nothing technically unlawful about parallel construction because it presumes defense attorneys will always know to ask about potentially unconstitutional techniques used in an investigation and be able to get that information during the disclosure process (a huge assumption, of course). As Human Rights Watch found in their exhaustive report, however, in practice parallel construction leads prosecutors to legally commit so-called *Brady* violations.[22] This occurs when prosecutors do not disclose to defense council evidence favorable to the defense's case (i.e., exculpatory evidence). If a defense attorney doesn't know to ask about potentially unconstitutional techniques (and why would they?), it is easy for prosecutors to just "not say anything" and let the case proceed (and why wouldn't they?). That *should be* illegal, according to human rights experts, but it isn't when done through parallel construction. It is hard to detect unless a defense attorney is incredibly persistent.[23]

Following the ruling that the spy plane was unconstitutional, it quickly became clear to me that police and prosecutors might try to use parallel construction. Instead, what I saw was prosecutors throwing out case after case before it even came close to a trial. Just as I was about to end my fieldwork, however, something changed.

Parallel Construction in the Novic Street Case

In 2023, I sat down with Detective Karim for one last discussion about the spy plane. After about an hour of conversation about cases that were dropped by the prosecutor's office, he dropped a bombshell: "Turns out we *are* going to prosecute one case [with spy plane evidence] after all. We decided we can let this one sneak through." I was shocked.

When he told me which case it was, I was even more surprised. The Novic Street case involved what the BPD called "supplemental surveillance." As discussed in chapter 6, in supplemental cases, PSS conducted tracking that violated the privacy policy advertised to the public. In the Novic Street case, detectives used the plane to track a red jeep associated with the suspect several days prior to the murder. This tracking unearthed several leads that detectives followed in an attempt to identify the suspect. The murder itself, however, was not captured by the plane, which was down due to bad weather. This is precisely the kind of search that the Fourth Circuit saw as problematic, from a privacy perspective, and got the program banned. How, I thought, could prosecutors possibly let this case, tainted by the most "poisonous" of spy plane evidence, go to trial?

In August 2023, I attended the Novic Street trial. Held over three days in one of Baltimore's creaky but still somehow regal courtrooms, I was excited because here, finally, would be my chance to see WAMI footage in an actual court proceeding.

Much to my surprise, however, the spy plane never made an appearance. It was not listed as part of the evidence in the court file. Its use was not disclosed to the defendant's attorney. When it came time to testify to his work, the detective didn't even mention that his team had made use of the plane, let alone use it in the very way that had led the program to be deemed unlawful. Detective Karim, who worked the aerial surveillance angle on the case, wasn't asked to testify. It's like the spy plane never happened. Instead, detectives presented the case using a "clean" set of evidence, which included CCTV footage, cell phone location data, and witness testimony. This would likely be a *Brady* violation in normal circumstances, but because it was done through parallel construction, it was considered okay. The suspect in this case, then, was *legally robbed* of his due process rights because police and prosecutors were too afraid to admit they used the spy plane.

Even though he didn't know about the spy plane's involvement, the judge was still skeptical of the investigation. He called the state's case "an interesting if unconnected series of events" and noted that there was "no identification [of the defendant] by direct testimony." After three days of presentation, the judge dismissed the case because it was just so incoherent that it didn't even deserve to be decided on by the jury. Even using the tactic of parallel construction to avoid disclosing the spy plane, then, fell flat. Prosecutors still couldn't put together a winning case.

The failure of this case did concrete damage to those who were involved. The victim's family never got an answer to who killed their loved one after being strung along through delay after delay. The case was dropped with the murderer still at large. The case also deeply impacted the life of the accused. After the trial, I asked the defense attorney if he knew that the spy plane had been used in the investigation of his client. "What spy plane?" he responded, confused. Not only did the attorney not know that his client had been surveilled by the plane, he didn't even know the program existed. After I gave a long explanation of WAMI and how it was used to investigate his client, I asked if knowing this information earlier would have changed anything about how he worked the case. He replied,

> If I had the opportunity at [my client's] bail review . . . to say, "Listen, Judge, we have credible information that, in the investigation of this case, the state used a spy plane . . . and all we know is, as of right now, as of today . . . it's illegal. It's a total violation of the Fourth Amendment." I would definitely have thrown that out there at any stage of the proceedings in an effort to try to benefit my client. But of course, if we don't know, we don't know.

Because of parallel construction, this attorney could not properly defend his client. Had he known about the program and its impact on the investigation, the case could have possibly been thrown out in the initial stages, during a bail hearing. "I mean, there's a violation here, for sure," the attorney reflected angrily. He explained that his client spent two years in jail awaiting trial, at which point he lost his job. He was then released but put on home detention for another year and required to wear an ankle monitor. He was not even allowed to look for work at this time, making him unable to financially support himself. Not only was he robbed of his due process rights, then, but he was also robbed of years of his life, placing a financial burden on his entire family. Once again, the experimental nature of the spy plane program created real concrete harms in the present.

The Problem of Objectivity

Like the spy plane, many crime-fighting technologies are presented by police to the public as not only better than but fundamentally different from human judgment. Humans are subjective. Machines are objective. Humans are biased. Machines are neutral. Human witnesses lie, misremember, and misperceive. Machines produce "raw" data, independent of human frailties. This is how the spy plane was portrayed to the public—an all-seeing eye that documents the "ground truth" of a crime scene, an "independent witness" for "both sides" of the courtroom. When you actually look at a technology like the spy plane in action, however, these binaries quickly break down. There is some sort of technical objectivity to the imagery, for sure—it "just" produces pixels that correspond to the material world. In order for those pixels to be at all meaningful or useful to people in the criminal legal system, however, they must be interpreted as evidence *of* something. What the spy plane teaches us about this interpretive process is just how brittle the mechanical objectivity of a technology can become. As soon as the pixels meet the eyeballs of an analyst, and that analyst passes their view down the chain of decision-makers, like a baton in a relay race, what seems to be "raw data" has actually become just another partial narrative, much like any other form of testimony.[24] The question, then, is why? What makes the objectivity of seemingly neutral, unbiased, and mechanical tools turn into brittle narrative evidence?

Part of the answer seems to lie in the structure of the organizational systems that surround these technologies, which function to make them useful for particular actors.[25] The spy plane, like a lot of experimental crime technology, is run through a public-private partnership, in which a for-profit contractor is tasked with providing an effective service to its main clients—police and prosecutors. This positioning can easily skew how the data are interpreted. Spy plane imagery is being made into evidence *for* a particular audience, so when the audience that controls the process is police and prosecutors, the data can be shaped to fit their point of view. Police and prosecutors can even make data disappear. There are mechanisms—*legal* mechanisms—like parallel construction that allow law enforcement to suppress unlawful techniques. Of course, as independent contractors, companies like PSS could always push back when they are asked to do dubious things. This is much easier said than done, however, when, as happened to PSS, your very livelihood as a company depends on being perceived as a good team player with police. The incentives of a public-private partnership make crime tech contractors structurally beholden to police, even though they are theoretically tasked with producing information for "both sides" of the courtroom.

I don't think structural biases tell the whole story, though. One of the things that surprised me about the way analysts, police, and attorneys handled the spy plane evidence was how much the complexity of the system seemed to lend itself to unethical and corrupt behavior. This behavior didn't seem to be driven by the nefarious intentions of "bad apple" police (at least not principally); rather, people were often just frustrated by how hard it was to get the spy plane system to work properly. There are so many things that have to go right for the imagery to appear self-evident and objective.

The cameras need to work flawlessly (during the day, in the right weather, in the right positioning, without obscurations, and so on). The analysts need to be certain about what the pixels really mean, detectives need to have an interpretation of the imagery that neatly conforms to their theory of the case, attorneys need certainty the evidence will be admissible in court, and everyone needs to be able to put on a convincing performance in front of a jury. When the stakes are as high as trying to solve a murder or determine guilt and innocence, the frustration of making this brittle system of people and machines work smoothly can easily lead to cutting corners. Unethical and corrupt behavior can be born of the defensiveness and desperation of working with glitchy, untested technology as much as it comes from ill intentions or moral failings.

This brings me back to the main point: the problem of experimentation. Whether it's the risk the BPD took in bringing an experimental tool into the courtroom, or it's the risk that individual investigators and attorneys took in twisting or hiding the spy plane evidence to keep a case moving forward, the people mainly holding the bag at the end of the day were poor Black residents. Those accused of a crime who wasted years in jail, victims' families who saw their hopes for justice dashed, the Black neighborhoods that felt the collateral damage of these failures—that's who shouldered the risk. Every time the brittle objectivity of the spy plane broke under intense scrutiny was an instance of experimentation. Surely there's a better way to protect public safety than this.

8   No to Hype, Yes to Community Control

It's the summer of 2023, almost three years since the experiment ended and two years since the program was deemed unconstitutional. As far as anyone knows, the spy plane is basically ancient history in Baltimore. I'm on the phone with Ross McNutt, who is sitting in the office space of the old AIR operations center, which PSS has hung on to all these years.

Unbeknownst to most Baltimoreans, McNutt is planning to fly a spy plane above the city again.

"We're going full Batman," he tells me. "What does that mean?" I ask, barely concealing my shock. "Well, is Batman dependent on the police?" he replies. "Oh," I say in astonishment. "We're in a totally independent capacity," McNutt continues. "No contract, no discussions, no interactions." I am stunned, but try to play it cool. "Wait, so how does this work?" I ask. McNutt explains,

> It's basically, we're working for the community. No contact with the police department.... We'll have our community advisory board to answer to, but this frees us up. We can publish [footage] on the Citizen app. We can publish it on YouTube. We're not limited. Tweeting.... We can watch the police, too. We think we'll be able to tell you where every single police car is every five minutes in the city.

Like any good start-up founder facing failure, McNutt has pivoted. PSS is planning to operate as a kind of free vigilante justice service. Rather than have to wait on the bumbling, slow, and corrupt police department to come save the city from gun violence, McNutt envisions he can strike out on his own, like Batman, and deliver justice straight to the people. Operating on behalf of an association of roughly fifty West Baltimore community members, with Simmons Memorial Baptist Church as its spiritual head, PSS plans to investigate crime with the plane and feed that information directly to the people who want it. There will be no public notification. No PR campaign. No democratic debate within the city council chambers. PSS has a plane, people want it, so they're going to fly it.

PSS is not alone in this approach. YouTube, X, Instagram, and other traditional social media outlets all have accounts dedicated to crimes in progress. Other start-ups are already operating in "Batman" mode, too, most notably the company Citizen (formerly called Vigilante). It provides a kind of virtual community watch service right on a smartphone. Download the app and people can post information about crimes they see outside their door. You can imagine the pitch to investors: "It's Facebook for crime." The company has even been known to hire civilians to become quasi-investigators, paying them twenty-five dollars an hour to live stream active crime scenes.[1] With the app already extremely popular in Baltimore, McNutt envisions a future for the spy plane that's like Citizen, in which PSS feeds information into this burgeoning online network of vigilantism.

Most importantly for PSS, operating in this space means the company is no longer beholden to the Fourth Amendment. "What happens if I'm not a government actor?" McNutt asks me rhetorically. "What limitations do I have as a private citizen, or as a company? I'm just providing security support to populations, right? Groups, businesses, downtown partnerships, companies—that sort of thing." This might seem strange and cutting edge, but, in a way, by going "full Batman" PSS is bringing high-tech surveillance in Baltimore full circle. Recall that the founding of the city's vast CCTV network, Video Patrol, was a downtown business association that put up cameras without any public debate. Of course, there is one major difference between this history and what McNutt is planning. The spy plane would be watching on behalf of poor Black citizens, not wealthy White businesspeople. What to make of this? Is this utopian innovation or dystopian decline?

We Should Stop the Boomer/Doomer Hype Cycle

The main argument I've been making is that debating disruptive crime technologies in terms of utopia or dystopia clouds our judgment. As historian of technology Lee Vinsel has pointed out, the main thing utopianism and dystopianism have in common, when it comes to technology, is hype.[2] Those who celebrate these new tools in a positive way—the boomers—put an almost magical faith in technological solutions, often stoking a false hope in citizens who are desperate for immediate help. Rarely do their utopian dreams come to pass. Citizens are strung along from failed program to failed program. Those who are afraid of technology—the doomers—also contribute to hype by concocting sometimes preposterous end-of-the-world scenarios. The reputations of the inventors who are the target of critique are actually bolstered by the negative attention. This "criti-hype," as Vinsel calls it, is sometimes an even bigger distraction because negative news has such high engagement in the digital media landscape.

The cost of letting the public conversation be dominated by boomer/doomer hype is that we end up debating the wrong things. Instead of talking about concrete harms in the present, we focus on some far-fetched tech future that, it is assumed, will occur unless we act now to prevent it.³ Or, if you're on the other side, we focus on some far-fetched tech future that *won't* come to pass unless we (and particularly government regulators) stand out of the way of the innovators (usually tech founders).

Increasingly, tech company executives have seen how they can actually be both boomers and doomers at the same time, and this creates effective cover for continuing to experiment on the public for profit. Nowhere has this been clearer than in the fantastically overhyped arena of so-called artificial intelligence (AI).⁴ In 2023, over 140 leading tech executives and computer science professors signed a one-sentence statement warning the public about the "existential risks" of AI for "all of humanity." Literally, their worry is that AI is going to lead to a mass extinction event. "Mitigating the risk of extinction from AI," the statement reads, "should be a global priority alongside other societal-scale risks such as pandemics and nuclear war."⁵ Signatories ranged from Bill Gates to Sam Altman (the head of OpenAI) to Chris Anderson (the "Dreamer-in-Chief" of the TED franchise). As Matteo Wong notes for the *Atlantic*, "The unstated assumption underlying the 'extinction' fear is that AI is destined to become terrifyingly capable. . . . The CEOs, like demigods, are wielding a technology as transformative as fire, electricity, nuclear fission, or a pandemic-inducing virus."⁶

It's telling that these signatories are expressing so much concern about the risks of AI now. They have largely ignored critics who have used concrete evidence to point out, for example, the already-existing harms of racial and class bias, mistakes, and overblown promises in many AI tools, especially those used in the criminal legal arena.⁷ But now that AI has started to bear fruit as an investment vehicle for venture capitalists, tech executives seem to have realized that the "doomer" framing can actually boost their profile. With AI talked about as a kind of unstoppable tidal wave of change, who else could save humanity from technology better than the very people who invented it? In this way, tech companies become both the source of and solution to the dys/utopian future, and in the meantime can steer attention away from some of the actual harms that are already taking place. When these technologies fail to deliver on their promises, and even their failures are not what we feared they would be, our attention shifts to the next big thing. Rinse. Repeat. How can we free ourselves from this endless cycle?

We Should Focus on the Problem of Experimentation

On the one hand, it makes sense that if we are ever going to get anywhere with making public safety interventions better (and I think we should!), we'll need to try new things. So, surely some experimentation will be needed. On the other hand, the case of the spy plane suggests that tech companies, police, and other law enforcement officials are the *last* people we should trust to conduct these experiments in a way that's good for the public.

Of course, I have been focused on just one unique case study. How do the harms from experimentation I saw in this case compare to what's going on elsewhere? What about other examples? These are really hard questions to answer, in part because there are so few instances in which we get to look behind the curtain of crime tech experiments. Still, even the small glimpses we've had of other technologies are revealing. As soon as you start to dig below the surface level of the boomer/doomer hype, the problem of experimentation rears its head, with eerie echoes of the spy plane.

Take PredPol, which had a long and lucrative contract with the Los Angeles Police Department (LAPD). This company uses algorithmic modeling of existing crime data to forecast future crime "hot spots," promising to help police more efficiently deploy scarce manpower. Critics have compared it to the dystopian film *Minority Report*, in which Tom Cruise, playing a detective, can see into the future and arrest people for crimes they have not yet committed. That's hype.[8] How does predictive policing actually work?

In a rare moment of transparency, the LAPD allowed sociologist Sarah Brayne behind the scenes to see how this technology was deployed.[9] She found that the forecasting was based on data that was unevenly concentrated in race-class subjugated communities, and thus tended to reinforce patterns of policing based on the racial and class segregation of Angelenos into neighborhoods that have been designated "high-crime areas" for generations.[10] PredPol's algorithm predicted that the LAPD would overpolice certain areas, and so they used the forecasting to justify their continued biased pattern of practice.

The LAPD canceled their contract with PredPol in 2020.[11] Though it is difficult to know for sure, this suggests that perhaps PredPol didn't add much. Indeed, one evaluation by the inspector general's office was unable to determine if the technology was effective at reducing crime or increasing clearance rates.[12] PredPol's contracts were dropped by several other agencies, too. PredPol didn't tell the police much more than they already knew, it seems, yet it fed racially biased crime statistics right back into the agency's deployment patterns, further exacerbating discriminatory policing. Garbage in, garbage out. Much like the spy plane, it sounds like PredPol may have become a force multiplier for racial segregation.

In the same study, Brayne found similar issues with Palantir, a digital platform used by the LAPD that links hundreds of siloed databases—from bank and ATM records to license plate readers—to crime records in agencies across the country. It's pitch: a kind of "Google search bar for crime." Similar to the spy plane, Palantir's main power, theoretically, is how it can stitch disparate surveillance data points together. Brayne found self-reinforcing data bias issues that were similar to PredPol's, but with the added problem that Palantir's software tools allow police to expand the dragnet of surveillance to the second and even third circles of association surrounding a suspect. Just like with the spy plane, the breadth of the system's gaze makes it difficult to differentiate between who is and is not "connected" to a crime. It brings people's mothers, friends, coworkers, and neighbors under increased police scrutiny just by being present in a geographic area at the wrong time.[13]

Okay, but does the technology help cops catch bad guys? It's unclear. There are no systematic studies that the public can access. Brayne found that detectives loved Palantir for the few great cases they could use it on, but couldn't tell her if it routinely added value.[14] So, Palantir helped detectives occasionally, but was/is it worth the cost? What kind of harms were unleashed by allowing police to use noncriminal databases to drag a suspect's entire social network into an investigation? If the spy plane has anything to teach us, it is highly likely that Palantir has led to a few truly spectacular case closures. It has probably "solved otherwise unsolvable crimes," but my guess is that this is an extremely rare event. The question, then, is does this offset the harms that the system created? Unfortunately, there are no systematic records of the system's harms with which we could debate Palantir in this way. No monitoring program was put in place. LAPD just "tried it out."

ShotSpotter, about which very little insider information is available, sounds like it might be even more of a pseudoscientific tool. Deploying sound sensors in strategic locations around a city, the company claims that its proprietary algorithm can differentiate a gunshot from similar sounds in near-real time and quickly send the precise location of the shot to an officer's cell phone. As discussed in chapter 4, Baltimore was a main test site for this technology and continues to use it. A recent evaluation of the program in Chicago showed that the technology's alerts are anything but perfect. Based on internal police data obtained through the Freedom of Information Act, the MacArthur Justice Center found that 89 percent of

ShotSpotter alerts in Chicago "turned up no gun-related crime" and 86 percent "led to no report of any crime at all."[15] In another study, the Chicago inspector general found that when police respond to a ShotSpotter alert, they are more likely to approach the scene amped up emotionally, and thus potentially injure or kill someone.[16] This may be why in 2021 a Chicago police officer "unnecessarily" shot thirteen-year-old Adam Toledo when he was found in the vicinity of a ShotSpotter alert.[17]

It gets worse. It turns out that ShotSpotter analysts actually have the power to override the algorithm and recategorize sounds as gunshots that weren't tagged that way by the computer. When the alerts made it into evidence in the prosecution phase, there was at least one documented case of police asking ShotSpotter analysts to alter the data for them.[18] When the algorithm categorized a sound as a firework, Chicago detectives asked ShotSpotter to manually override the alert to recategorize it as a gunshot *and change the location of the sound* to better fit the other evidence detectives had in a homicide investigation. This sounds a lot like what I saw when detectives and attorneys led PSS analysts to manipulate their interpretations of spy plane footage to fit different agendas. ShotSpotter data may be extremely brittle, causing officers to, perhaps out of desperation, engage in unethical and corrupt behavior. It is perhaps no surprise, then, that Chicago's newly elected progressive mayor, Brandon Johnson, cancelled the city's multimillion-dollar contract with ShotSpotter in 2024, following a long grassroots campaign by activists who saw the technology as a dangerous experiment gone wrong.[19]

What about facial recognition technology—what many doomers see as the leading indicator of society careening toward dystopia and boomers see as a new breakthrough for cop vision?[20] Dozens of companies, such as ClearView AI, have sprouted up hyping their ability to almost magically identify a person based on a facial image pulled from a security camera, cell phone, mug shot database, license photo database, or even a trove of social media images. Feed the algorithm a "probe photo" and—bing!—it returns the identity of the person.

Engineers inside the machine learning community have been sounding the alarm for years about how flawed these tools can be in practice, particularly when used on Black, Brown, and gender nonconforming people.[21] Detroit, for example, has been trialing real-time facial recognition on its CCTV cameras in a small, race-class subjugated neighborhood. Known as Project Green Light, it is unclear how effective the program has been at reducing crime, but it has already made some headline-grabbing mistakes. In 2018, Robert Williams was falsely identified as a suspect in a shoplifting investigation because police fed a poorly lit image from CCTV, which did not even capture the front of the person's face, into the facial recognition algorithm. Being a machine, the algorithm returned several hits anyway, one of which was Williams. Williams was arrested in his driveway, in front of his wife and kids, and held in a detention center for thirty hours before the mistake was detected. He sued the department.[22] Much like with the spy plane, it seems that facial recognition might have a false positives problem that is racially biased.[23]

The NYPD's (mis)uses of facial recognition are even more problematic. Documented most extensively by the activist organization Surveillance Technology Oversight Program and the Center on Privacy and Technology at Georgetown Law, officers are known to have doctored probe photos of suspects in Photoshop in order to make them easier to read by the algorithm. Police have actually used these doctored photos to effect arrests.[24] In these cases, police are often using facial recognition under less-than-ideal circumstances (poorly lit footage, for example) and in entirely different ways than the companies intended.

One explanation for why they engage in this kind of abuse is that police are ill-intentioned, some might even use the word *evil*, especially when it comes to investigating crime in Black and Brown neighborhoods. That could very well be true in some instances. My experiences with the spy plane suggest a parallel explanation, however. My guess is that, like the spy plane, tools like facial recognition are so complex, glitchy, and fragile that police get frustrated, defensive, and desperate. The algorithm is brittle. One tiny thing goes wrong and the tool doesn't work. Meanwhile, the pressure is on to put up good arrest and case closure numbers. Now the tool they were given that is supposed to make them supercops has actually become a liability. So, they start to cut corners in order to make the technology "work," even if this means making unethical decisions or engaging in corruption.[25] This is no less reprehensible than outright ill-intentioned behavior, but it is also a possible structural explanation for chronic abuse that can't be reduced to a simple story of "bad apple" officers.

Based on what I saw with the spy plane, my hunch is that police entered into these contracts expecting the technologies to just "work" like the powerful objective tool they were sold. They expected a scientific instrument and all they got was a pseudoscientific prototype. They quickly found that there were few instances where the tool could be used exactly how the company described. The real world is so much messier than the company's brochure. So, out of frustration and desperation, police take the role of mad scientists. They tinker with the technology, using it in other ways because . . . well . . . it's here, and it's being paid for, and so "we might as well use it." Police then end up using it in ways the company itself probably hadn't thought of, or might even object to, resulting in all sorts of unforeseen risks and harms. As sociologists have been saying for nearly a century, technology almost never works perfectly and is almost always used in ways designers didn't intend.[26] In the end, police, city officials, and tech companies have to scramble to hide these shortcomings and mistakes in order to keep up the impression that they are operating ethically. This kind of experimentation, I suggest, is a breeding ground for corruption.

If I'm right that police often misuse and abuse these new technologies because they are experimental, do the companies push back? When they see that police are using their tool in ways that weren't intended or are even unlawful, what do they say? In most cases, we have no idea because there was no one behind the scenes to see it. Knowing what I know about how things played out with the spy plane, my guess is that companies don't often push back because of their for-profit arrangement with law enforcement. If police are the main client you need to please, and your company's future depends on delivering customer satisfaction, these companies start to function more like a help desk than a truly neutral, independent witness. Again, the problem may be structural.[27] Companies surely don't want police to misuse their tools, but they also probably feel like there is little they can do about it without sacrificing a long-term contract with a big-name client.

To sum up, new crime technologies are not only glitchy and underperforming, but flawed in systematically biased ways that users seem unable to anticipate. Through the process of experimentation with imperfect technology, then, harms keep popping up at various points: in the technology itself (i.e., a biased algorithm), in the way the technology is intentionally misused by police, in the way user error can skew information in patterned ways, in the way forensic experts and attorneys misinterpret or even willfully manipulate the technology as evidence in court, and many other points besides, I'm sure. This doesn't sound like some far-fetched dystopian future that we need to brace ourselves for with the help of our techno-solutionist saviors; it sounds like regular people who've been given free license to tinker with an untested tool that hurts people. That's our present. It's not good, but it's also not dystopia.

Experimentation Refocuses the Conversation

The main payoff of centering the problem of experimentation is that it helps us talk concretely about racial and class inequalities in a way that the boomer/doomer framing often makes us miss. Both utopians and dystopians are focused on the implications of a tool that can "watch us all." Not only is this rarely even close to the true scope of a technology's power, but even when a technology *becomes* considerably more powerful than what came before, it is often first used within a small subset of race-class subjugated communities to discover that power. Black and Brown communities are used to test and refine it.[28] So, even if "one day" a technology comes to be an existential threat, it has already long been a threat to subjugated communities. Wider society just ignored the warnings.

Focusing the debate on experimentation can help bring these warnings to the foreground sooner. We stop debating how good or bad the technology is theoretically, in terms of some future state, and start debating if it is really plausible that it can be deployed in a way that does the most good for those who have the most to lose, without also causing too much harm. That very well may mean we simply don't deploy many experimental crime technologies because risk and reward are so unbalanced and unjust. On the other hand, experimentation, if done right, could very well give community members a level of control that contributes to a more democratic and transparent way of doing public safety. We just need a way for public officials and community leaders to differentiate between good and bad experimentation. That, as I explore next, is a better starting point for the conversation.

We Should Regulate Experimentation

It is uncontroversial to say Baltimore and other cities struggling with gun violence need to change the way public safety is done. The level of violence is unspeakable. That means we need to try new things, and doing so will require something like experimentation. What's the best way to do this? Is there a way to deploy new technologies, perhaps in conjunction with non-tech-centric interventions, in real-world scenarios that is better than the way it is typically done?

There are two ways we could go: regulate public-private partnerships as they are currently designed, or create new ways of organizing public-private partnerships that will avoid the worst and harness the best outcomes. Fortunately, we have many good examples of both pathways for experimentation.

There has been a lot of thought already about how to regulate standard public-private partnerships in policing. As Sarah Brayne argues, first and foremost, we need to *slow down*.[29] There should never be an instance in which an untested tool, with no independently gathered information about if or even *how* it works, is deployed on the public. "Rapid prototyping," "fast failure," and "fake it 'til you make it" *might* be excusable if a start-up is looking to develop a slick new app, but it is a completely inappropriate philosophy for people with a badge and a gun. It has no place in the criminal legal system and should be banned. There is simply too much at stake if the experiment fails or, even worse, harms someone.

Instead, we need to approach experimentation with deliberate care. Most importantly, the deployment of a new technology on the public should happen only if there is already independently gathered evidence that answers basic questions: How does the technology actually work in practice? Is it reliable? What are examples of biases, glitches, or malfunctions that haven't yet been solved in the development phase? What kinds of harms could these unaddressed issues produce in the deployment phase? How do the harms stack up against the potential benefits, if the technology improves through experimentation? What do success and failure look like? Who gets to define them?

This information needs to be *independently gathered*. If, as PSS did, a tech company comes to public safety officials with an internally conducted study, a study funded by investors, or a bunch of focus groups conducted by the head of the company, that's not good enough.[30] It can cause the public to appear to be in favor of an experimental tool, but their opinion is based on incomplete or slanted information. Citizens need to be allowed to build their opinions from information that is untainted by political and financial interests.

Put succinctly, public-private partnerships need more publicness and less privateness. A good model might be an impact assessment.[31] These kinds of evaluations can bring engineers together with social scientists, public health officials, and local residents with close knowledge of the context in which the intervention will happen to better understand its impact on specific people in specific places. The risks of deploying the spy plane in St. Louis, for example, are different from those in Baltimore because the two cities are geographically different and have different deployment patterns of other surveillance technologies. The risks of using the spy plane in North Baltimore are different from those in West Baltimore because of the history of segregation. If a tech company comes to the table without this kind of knowledge of a city's unique geography and history, that should be a red flag. Officials should tell the company to come back when they are better prepared, or just go away. Cities have different local dynamics. These kinds of human and social factors should be taken into account *before* a tool is deployed, not after the impacts have been discovered when they show up as mistakes, biases, and harms. This means that we desperately need specialists other than tech founders, venture capitalists, engineers, and police experts to shepherd a new tool into the deployment phase. We need a whole-community approach.

Another aspect of what I mean by more publicness and less privateness is *transparency* and *accountability*. It should never be the case that police get to try out a new tool without having to report back to the public how the experiment is going *as it is happening*. The spy plane is a telling example in this regard. On the surface, the BPD had transparency and accountability systems in place. There were three teams of academics assessing the program from the outside. Yet, none of them discovered some of the most egregious harms of the spy plane, such as mistaken tracking and false positives. The BPD also released a couple of interim reports, but they used these reports to hide some of the most important details about the program, such as whether it had actually resulted in any meaningful arrests. Ongoing information about the program was tightly controlled by police, always with an eye toward PR. Once the program was shuttered, moreover, there was never public dissemination of information about how the people who were arrested with the plane, victims, and their families fared when the cases went to court. These were some of the most meaningful and harmful outcomes of the experiment for the public. These kinds of key issues were not systematically documented, in part, because (other than a lone sociologist) no one but police, prosecutors, and company insiders knew what was going on. There was no regulatory framework in place requiring officials to track these risks.

The lesson here is that independent evaluators, rather than just being given information through a request for documents, such as an internal police report, must also be *closely embedded* with analysts, police, and attorneys. As I saw with the spy plane, evaluators might not know what information to request unless they are on the scene to watch the problems unfold. Police and prosecutors are also really good at not recording, "forgetting," or even manipulating data to hide the way they've cut corners or made mistakes. Most importantly, then, police and their private contractors should be required to submit to close inspection that can systematically document the experiment *as it occurs*. There should be consequences when those harms are revealed. For example, someone other than police or the contractor should have the authority to pause or pull the plug on an experiment if harms arise. Going further, when cases that involved the new tool make it to the prosecution phase, the public should get to know how the technology performs in court.

This style of regulation, transparency, and accountability will mean developing a democratically run community oversight program with real teeth. To be truly effective, it should be funded separately from the sources funding the experiment and use evaluators that are not affiliated with the tech company or the criminal legal system. It should also have more than a consultation function, but have real power to hire and fire personnel, and terminate programs in progress. Again, local community leaders, public health officials, and social scientists might be an excellent source of expertise for this kind of oversight function.

Envisioning and implementing these kinds of community-led regulatory systems will be really difficult. It will mean a long fight inside (often pro-police) city governments and against the entrenched culture of police departments to operate with total discretion and superprivacy. Oftentimes, these kinds of regulatory reforms are done piecemeal, city by city. Each group of reformers must start from scratch. A more overarching national framework would be helpful. The ACLU, for example, has begun to work toward this. Its Community Control Over Police Surveillance (CCOPS) model legislation is one of the first concrete programs for helping reformers quickly draft more robust and democratic oversight programs.[32] More efforts like this are needed.

We Should Experiment... Better

The biggest problem with a regulation approach to experimentation is that people living amid constant gun violence need safety, and they need it now. Telling police and community leaders to slow down and do things more carefully might create good long-term outcomes, but it could also get so bogged down in politics that no changes end up taking place. Or, at best, changes happen slowly while many people continue to suffer from unrelenting violence. If that happens, people are likely to give up on reform, and things will just go back to the old ways.

There needs to be some creative, even rapid, experimentation that does things differently from the usual public-private partnership. These kinds of experiments will be really risky, but perhaps there is a better way to do it with and for the communities that have the most to lose *and gain* from them.

The most promising experiments I see are happening parallel to or even entirely outside the traditional criminal legal system, within public health approaches to violence reduction.[33] Violence interruption, psychological-behavioral interventions, and focused deterrence models are the most well-known experiments. Unlike traditional policing, which often treats violence as individual immoral behavior, this paradigm sees violence as a kind of contagious virus that spreads through a community's social network. As with a virus, we can use targeted health interventions to stop the spread.[34] Some of these experiments even involve the use of cutting-edge technologies, like automated decision-making.[35] What differentiates these from the mainstream, however, is that they are more like public-public partnerships, often bringing together community groups and academic institutions, cutting out for-profit companies entirely or subsuming them to a merely technical role of providing and operating a tool. Right away, then, these programs remove what is an obvious perverse incentive: profit.

A good example is project READI in Chicago. Standing for Rapid Employment And Development Initiative, this program starts from the now well-founded principle that the majority of gun violence in most cities is concentrated in a small social network, whose members are often both perpetrators and victims.[36] READI aims to identify these "high risk" individuals and rapidly provide them with a combination of cognitive-behavioral therapy (CBT) and steady employment to disrupt the cycle of violence. While this sounds simple, identifying exactly who could benefit the most and (even more difficult) where they are is surprisingly hard. It requires investigatory work that ultimately looks a lot like location tracking.

For READI, there are both similarities to and differences from traditional policing in how this tracking is conducted, as well as the end goal at which it is aimed. The program tracked down high-risk individuals, in part, by working with local community organizations, rather than police. Their outreach workers have strong bonds of trust within their neighborhoods. They are often formerly incarcerated and have deep knowledge of how violence operates in their local social network. They know what conflicts are about to bubble over into violence—something police rarely understand. These community leaders patrol and watch their neighborhoods, investigate violence, and follow individuals targeted for intervention by the program. In this way, there is a passing similarity to detective work.[37]

READI also used technology to assist these outreach workers. They deployed an algorithmic risk assessment tool to analyze municipal data and identify high-risk individuals that were not already known to outreach workers. While these kinds of tools are now common in police departments, they are usually used to map gang affiliations or mark individuals for extended surveillance, with an end goal of arrest and punishment.

Both of READI's monitoring techniques are different from traditional surveillance in one key way: instead of punishment, READI delivers services that, research shows, can help pull someone out of the cycle of gun violence. Is this surveillance? Well, sort of. It is certainly "veillance," given that both humans and machines are watching and keeping tabs on target individuals, but it is not quite "sur." What is being delivered is not punishment from above but an offer of assistance to get one's life back on track. It is perhaps a kind of "lateral surveillance" or "peer monitoring," in which the watchers are trying to place themselves as support services alongside the watched.[38] The watched ultimately have the final say to opt in or out of the intervention.[39]

While the effectiveness of READI Chicago is still being studied, the program has shown some promising, if inconsistent, results in a mid-trial assessment.[40] Importantly, however, it has been careful in the way it has experimented and has been transparent in reporting how things are going, *while still acting swiftly*. Care and speed do not have to be antithetical. READI partnered with a team of independent academic researchers to evaluate the program as a randomized controlled trial. The researchers used a lottery system to randomly select individuals into the intervention, thus allowing them to make apples-to-apples comparisons between those who were offered READI and those who weren't. When the final results come in, evaluators will know with some certainty if READI had an impact that is independent of other factors.

Second, the evaluators specified *before* experimentation what the effectiveness of the intervention means so that they could not later cherry-pick the results and tailor them to fit the program's PR narrative.

Third, the very premise of the intervention is based on some prior evidence that it could work. Both CBT and employment have been shown in other research to reduce the likelihood of becoming involved in gun violence; they just haven't been used together and through this particular targeted, peer-monitoring delivery mechanism.[41]

Finally, independent external evaluators are embedded inside the experiment, gathering real-time qualitative data—interviews, observations, surveys, and focus groups—to understand what the outcomes, good or bad, actually *mean* in the context of participants' lives.

READI is also not perfect. It's not a utopia. While I applaud the structure and organization of the program, it is still an experiment. Like the spy plane and other criminal legal experiments I have documented in this book, READI is using race-class subjugated communities as a petri dish. It thus increases the risk profile for folks who are already vulnerable, especially if it turns out not to work. The community outreach workers who are at the heart of the program, for starters, are being asked to intervene in incredibly dangerous situations. These workers are rarely given psychological services to help them cope with the trauma of providing care in such volatile circumstances. Outreach workers from a public health–based program in Baltimore, for example, have been shot and killed on the job.[42] The pay, if they get it, is often not that great.[43] Political winds also shift wildly around experimental programs like these, which cannot depend on the reliably large budget of a police agency. A change in city leadership can mean that outreach workers are suddenly left without a job after having spent years building trust with community members.[44] What will happen to them, and these fragile relationships, when the experiment is over? What harms could be unleashed if these relationships are suddenly broken apart?

For the recipients of READI's interventions, the stakes are also incredibly high. What will happen to the participants if the program is found not to be effective? Could there be unintended harms by exposing people to CBT and a job that, if the recipient continues to be drawn into the cycle of gun violence after the program, actually make the situation worse? Could they become an even bigger target for retributive violence once there are no longer READI workers looking out for them? What if the algorithmic risk assessment tool that identified a person as "high risk" made a false positive? What kind of message does it send to a person when they were targeted as potentially suffering from the "disease" of violence? Will it cause them to distrust and retreat from society more?[45] Is a randomized controlled trial really an ethical context in which to engage in this kind of targeting for "treatment"?

The language of medicine—interventions, outcomes, violence "contagion"—might sound less threatening than law enforcement language, but the racial history of the medical profession is just as horrifying as the history of policing.[46] A great next step for public health approaches, I suggest, would be a deeper interrogation of the racial meanings of violence reduction as medicine. If communities are going to run these experiments, and I think they should, more consideration of these ethical questions is needed. On the other hand, traditional police-controlled experiments are rarely held to this high of an ethical standard, so why should we make this more community-controlled model bear a double burden?

Full Batman

As he has done so many times since I've known him, McNutt shocks me again. He tells me that he has a new camera he's planning to try out in Baltimore. He has designed a 150-billion (with a *b*) pixel system that will "walk over" the city in an orbital pattern, taking dozens of extremely high resolution photos of slightly smaller sections of the city than the WAMI cameras. This camera will sit alongside the WAMI system on the spy plane. Each orbit of the plane will create constant, second-by-second WAMI photos, as well as a mosaic of slightly smaller photos at much higher resolution. Analysts can track with WAMI until the target enters one of these higher resolution patches, switch to the new system, and identify basic details about people's clothing color and vehicle attributes—all from ten thousand feet. No police or ground camera infrastructure involved. "We can't yet identify a license plate [with the new camera] from that altitude," McNutt says, "but we're working on it." They'll use the Citizen app, rather than police dispatchers, he tells me, to identify crimes in progress, and then immediately be able to post to the app that PSS has imagery available. Then, they can just work directly with the folks involved, offering up leads. Full Batman.

The way he plans to fund the service is even stranger: the Crime Stoppers program. With branches throughout the country, Crime Stoppers is a nonprofit organization that provides rewards for anonymous tips that lead to the arrest of criminal suspects and fugitives. In Maryland, they offer $2,000, in cash, for tips on crimes ranging from auto theft to murder.[47] McNutt told me that "we see so many hit-and-runs and carjackings in our images, we could probably fund the entire program just off that." In other words, PSS plans to use the spy plane's relatively higher rate of success at solving property crime to create a constant funding stream from Crime Stoppers money.[48] This will allow them to provide analysis of murders and shootings for free to community members.

If this new version of the program were to be implemented, it would eventually break in the news. It is highly likely that it would then get roped into the same worn out boomer/doomer hype cycle. Let's not. What if we looked at it through the lens of experimentation? On the one hand, there are some good things about this experiment. It centers the needs of a certain segment of West Baltimoreans, placing them in an advisory capacity to hold McNutt's team accountable. That's promising. The new camera system, though it is obviously incredibly invasive, might help analysts avoid some of the troubling false positives that I have documented in this book. That could actually be a good thing. Even if the evidence from this version of the spy plane doesn't end up providing conventional justice inside a courtroom through the work of detectives and lawyers, it could still potentially provide basic information to victims' families. That's certainly better than what they've got. Vigilantism, as undemocratic and opaque as it is, could provide something important when the conventional investigatory system has become so broken: answers. How could we blame a grieving mother for going straight to PSS for answers if she could get them? That's what I would do, too.

On the other hand, this version of the experiment seems even less transparent, democratic, and careful than before. Though a group of West Baltimoreans are in an advisory capacity, they don't represent all the Black Butterfly. The Black community is not a monolith. There are many different opinions within West and East Baltimore about how the spy plane should operate (and a vocal minority that thinks it should just go away). There is no democratic process in place to ensure those voices are heard. There is also no independent, external team of evaluators to keep tabs on the operation, measure its effectiveness, and document harms. How will we know to pull the plug on this experiment before things go spectacularly wrong? Who gets to say when to stop? PSS claims that this new camera system will fill the gaps between the wide angle and ground level, but does it work reliably? There has been no impact assessment. Has it been tested enough to be deployed in such a high-risk situation? What kind of unforeseen harms might this new camera cause in different parts of the city? Lastly, I can envision all kinds of racial bias and retributive violence coming from this program if it is relying on the Citizen app. Citizens can be just as racist toward their neighbors as police. Won't people just use the plane to take the law into their own hands? Is it possible that people could fool PSS into helping them track an ex-lover or other person they wish to harm? In the end, it still feels like PSS is using Black lives as test subjects, with little transparency and now even less democratic accountability. My hunch is that *because* of the problems of experimentation, there will be things that *both* the boomers and doomers find alarming. They have more in common than they think.

What strikes me about our contemporary moment, standing in the fading light of the techno-optimism of the 1990s and propelled into a dark future of profit-driven techno-solutionism, is how difficult we find it to talk about technology as a fundamentally human thing. Citizens, journalists, politicians, and other observers need to stop getting baited into a debate about some far-off tech future in which a tidal wave of magical gadgets seemingly autonomously overpowers our puny human institutions. This is not only unrealistic, but turns out to provide great cover for those who stand to make spectacular profits from our naivety about what's really going on behind the scenes. Technology is not god or the devil. Technology is people. It cannot save us any more than we can save ourselves.

APPENDIX
Watching the Watchers

Baltimore is a second home to me. Though I grew up in an upper-middle-class family in the American Midwest, my partner's family has lived in the city and its northern suburbs for three generations. Like a lot of White families, they have mostly lived in the White L, the city's narrow central corridor, venturing into the inner harbor for fun. When I first married into the family, this was my impression of Baltimore. In my early twenties, long before I called myself a sociologist, my partner and I helped run a summer camp that brought West Baltimore kids out to the countryside, beyond the northern suburbs, to enjoy the forests and streams, play games, and hang out. We would sometimes travel back with the kids to their neighborhoods in the city, crossing the lines of segregation that had filtered my perception up until that point. Like a lot of sheltered White kids from the Midwest, I had no idea just how dramatic these lines could be. As sociologist W. E. B. Du Bois observes, White Americans live on the other side of a racial veil, often blissfully ignorant of how Black folks must cross it every day.[1] When White people do cross the veil, it is often by choice, and it might even be experienced as exhilarating and exotic. That's how it felt to me back then, if I'm honest.

Some decades later, now a professional sociologist, I embarked on studying the spy plane with an acute awareness of this classed and racialized relationship to the city. I knew how much my identity would probably play on my perception. I made a commitment from the outset, then, to try to use my race and class identities to, as anthropologists like to say, "study up."[2] Rather than attempt (and most likely fail) to articulate what surveillance is like on the ground from the point of view of the watched, I figured I was much better placed to study the watchers—those higher up in the power structure who are often protected by their superprivacy from public scrutiny.[3] I can blend in with cops pretty well. I know how to engage in the kind of banter that cops do. I grew up playing sports. A lot of my friends from back home went into the military. Throughout the research process, I relied heavily on these aspects of my identity to gain and maintain access.

Relationships with people in PSS and the BPD proved hard to navigate because of this orientation. I did not want to present myself in a disingenuous way. I am skeptical and critical of traditional policing. I had serious reservations about surveillance technologies like the spy plane going into the project. I wanted McNutt and others to know that I would write critical things, if I saw things to critique. On the other hand, I knew that my access to the program was unusually privileged and could crumble if PSS personnel and police didn't trust me.

This tension came to a head when PSS's lawyer dragged me into his office on one of the first days of fieldwork in the operations center. He told me that I needed to sign a legal contract with the company if I was going to be there. Having spent three years gaining enough trust within PSS to be allowed in the door, having literally just taken out my field notebook, I was being told that I was not authorized to look at anything on a screen. I was gutted. Fortunately for me, the Arnold Foundation, which put up the money for the spy plane trial, had already hired scholars from other academic institutions to conduct external evaluations. So, PSS crafted a legal document, based on those already used by the Arnolds, which designated me a "consultant" loosely attached to PSS. The agreement tasked me with conducting an internal audit of the company's analysis service. I was thus able to get a security clearance through the BPD, a key card with full access to the spy plane data, and a certificate from the police commissioner giving me permission to be out of quarantine during the pandemic "lockdown." After three days of back and forth with the company lawyer, PSS even agreed to insert the clause "total academic freedom" into my contract. The company agreed to let me do my research without interference and to own the work product.

As part of this agreement, McNutt asked that I help out with small research tasks that might improve the company's operations. I did a couple of things. I helped them craft an internal customer satisfaction survey for detectives. I reviewed cases where detectives never even opened PSS's evidence packets to see why. As I discuss more below, I also engaged with the NYU Policing Project, one of the teams of external evaluators, on behalf of the company to help orient their work. Though this meant I would be assisting the company in some way, then, it seemed like a fair deal in exchange for the level of access I was being granted.

PSS also offered to help me access research funding through the Arnold Foundation, which had been set aside for the external evaluation process. This is where I drew a line. Something in my gut told me that taking money from the Arnolds would muddy my perspective and perhaps sully my credibility, so I turned down the offer. The research for this book, then, was funded entirely by Williams College. Knowing what I know now about the problem of experimentation, I have come to appreciate the importance of financially independent auditors even more.

Watching the Watched

Despite my attempts to keep my analytical gaze trained on the watchers, it was only a matter of time before I would need to talk about the watched, too. At various points in the research and writing, I found that I needed to gather and share information about the experiences of citizens in Baltimore's race-class subjugated communities who were investigated by PSS. How else could I show the human impact of experimentation without sharing these stories? For me, this was the most ethically challenging part of the project. What does it mean for me—an upper-middle-class, White, nonlocal academic—to share stories about the investigation, murder, and trauma of Black Baltimoreans I don't know? Black death in Baltimore has long been a source of exotic titillation for White consumers through the media, such as the popular cop drama *The Wire*. I was worried about reproducing this harm.

I did all the basic things that ethnographers do to protect the confidentiality of the people who appear in the research. I used pseudonyms and changed key details about the cases I analyze so that it would be extremely difficult to figure out the names of the people involved. I protected my field notes and didn't share confidential information with anyone beyond those who were there. I still felt like this wasn't enough, though. I felt like I really needed *permission* from those who were directly affected by spy plane investigations to share their stories.

Because of my race and class identities, however, and because of where I was positioned within the operations center, making authentic connections with suspects and victims' families was extremely difficult and possibly unethical. This became especially complex when it came to discussing the experiences of victims' families. On the one hand, it seemed wrong for me to simply ignore the concrete harms the spy plane created for these families by leaving their stories out of the book. Of all the people affected by the experiment, it is those who had closure snatched from them by failed prosecutions that suffered the worst. On the other hand, it seemed unethical to ask families (especially poor Black families) to share their trauma for an academic text and to write about one of the most personal things that can happen to a person—the murder of a family member—without their full consent and opportunity to contribute or opt out. Ethnography has a long history of doing this, due in large part to its colonial past.

I decided to reach out to multiple people knowledgeable about gun violence trauma in Baltimore for advice, including several Black West Baltimore community organizers and a local investigative journalist who has worked on Baltimore crime stories with a high level of sensitivity and care. They all said the same thing. These families are grieving. Inviting them to share their stories in the format of an academic or journalistic interview with a White researcher who has been working alongside police would be retraumatizing. They also said that approaching the families in person or by telephone would be perceived as too aggressive and only raise suspicions that I was working *for* police. On the other hand, they agreed that including the stories was essential and that I needed to at least notify the families of what I was doing and give them the opportunity to respond. They advised I reach out by letter to the families of the victims in the three murder cases that I discuss most extensively in the book. This would be less aggressive and give them more control.

In the letters, which were handwritten on regular notepaper rather than typed on formal letterhead, I explained how I would discuss the *investigation* of their loved one's case, protecting the confidentiality of their loved one as much as possible by changing key details. I invited family members to contact me if they wanted to share anything more. I also reassured them that I would adhere to their wishes if they did not want the information in the book. I also included a formal business card with contact information and a link to my bio.

I did not receive a reply from the families. Having no explicit permission to tell these stories, but also feeling an obligation to show the documented harms of the spy plane program, I ultimately decided to include them in the book.

When it came time to actually write about these experiences, though, I made some careful stylistic choices. I decided I would try to write about them in a way that would not play into the tropes of exotic titillation that characterize so much "true crime" entertainment. In the opening vignette of the book, for example, I describe a cell phone video obtained by detectives of a horrific daytime murder. This video was made by one of the victim's family members and shared with PSS analysts via the detective (already ethically dubious). The video was not directly shared with me for the purposes of a book. In an early version of the manuscript, I wrote about some of the gruesome details in this video because I wanted to give a full accounting of the kinds of evidence analysts were working with as they did investigations. Workshopping the chapter with other scholars made me realize that this level of detail was not only unnecessary for my analysis, but played into damaging tropes where Black death stories are extracted for White audiences' entertainment. On revision, I took out most of the details about the video, leaving only the parts necessary to focus attention on what investigators did with it. To be clear, some readers of my earlier draft *liked* reading the details about the video. That's precisely why I took them out.

In another instance, I decided to share more details about the experiences of a victim's family, rather than edit them out. One of the main complaints about the BPD, like other agencies in high-crime cities, is its abysmal case closure rate for violent crime. While the statistics around this rate are regularly discussed in the media, rarely do the media discuss the impact a failed prosecution has on families. Having closely followed some the spy plane cases, I was made privy to stories about how families reacted when they received the news, often years after the death of their loved one, that a case had been thrown out. Though I did not have permission from the families themselves to share these secondhand stories, the information just seemed too important for documenting harms that almost never get discussed. Moreover, they are almost never discussed *because* of the racial and class identities of those who are being experimented on. The sentiment behind so many criminal justice experiments is, if the program fails, who cares what happened to the test subjects. In this case, then, relating these stories, despite not having full permission, seemed not only "on point" to the argument of the book but also important for creating a human story where there are usually only cold statistics. Still, I acknowledge and regret that there remains something extractive about conveying this information without full communication with the family.

I still don't think the way I navigated the race-class dimensions of this project was perfect. I did my best, but ultimately there are aspects of the ethnographic process that are, by nature, extractive and exoticizing. I simply could not figure out ways to do the project without somehow playing into these harms, to some extent. Shy of not doing the project at all—and maybe that's the only pure ethical choice—the best I felt I could do was lessen the harms as much as possible by being hyperaware of my positionality and carefully balancing visibility, invisibility, and consent.

Affecting the Outcome

Given my level of access, including seeing investigations in real time, there were multiple occasions when I realized that I had the power to affect the direction of the spy plane program. Qualitative social researchers and their critics have long grappled with the question of whether and how an ethnographer's presence shapes the data.⁴ But this was on another level. This project is a study of an ongoing event—an experimental trial—and so my presence could not only shape what people said and did in front of me, but also the very course of the experiment itself. Here's how I handled those situations.

The core activity of my fieldwork involved sitting next to analysts as they tracked. It can be incredibly tedious work, so having an extra pair of eyes around is always helpful. I made a decision in advance of the fieldwork to never assist analysts in tracking.⁵ This was partly an ethical decision. Though I had made a commitment to "studying up," it did not mean that I needed to "go native" with the watchers and become a full participant, contributing labor to the business of putting people in prison. I decided that the best position for me was to take a strictly observational role. I quickly found that this was not only safer ethically but also legally. The rule in Baltimore courts is that any analyst who even touches an investigation, whose key card number is registered in the tracking logs of a case, can be subpoenaed as an expert. In this sense, being a full participant would have put me in the uncomfortable position of having to testify in a criminal trial. In short, though I could have, I chose not to affect the outcome of individual investigations.

The things I was seeing behind the scenes were also germane to another set of legal issues: the civil case against the BPD. When the ACLU helped Leaders of a Beautiful Struggle and other appellants sue the BPD over the spy plane, I realized that my research might prove fruitful to their efforts. I saw that I could, at least theoretically, reach out directly to the ACLU and offer assistance in crafting their legal strategy in the case against the spy plane.

Before I could seriously entertain that thought, however, I was contacted by researchers from the NYU Policing Project, who had been tasked by the Arnold Foundation to conduct an ethics and privacy audit. Originally, this team was meant to be on-site. McNutt had even begun renovating an office next door to the operations center to provide them with space to conduct their research. Because of the pandemic, however, the NYU team could never come down. They contacted me to help them be their eyes and ears. "What are you seeing?" they asked me. "What should we be focusing on?"

What role should an ethnographer play in this situation? Are ethnographers distant, objective scientists who should never get so close to their subject that they muddy the results? Or should ethnographers be more like investigative journalists, who are there to expose the truth in order to hold those in power accountable? If I took the distant scientist approach, I surmised, I would say no to the NYU team because it would affect the spy plane experiment, the very thing I was studying. I should really let them do their work as though I never even existed. If I took the journalistic exposé approach, I would start feeding the NYU team (and probably some real investigative journalists, too) regular dispatches from the field.

To be completely honest, I did not know what to do here, and I ended up acting on instinct. I hedged. I told the NYU team, "If it were me, I would start taking a look at what the company calls 'supplementals.'" As discussed in chapter 6, these are cases where analysts conducted surveillance that not only broke promises made in the MOU with the city, but violated Fourth Amendment law. I didn't tell the NYU team my legal reasoning for singling out supplementals, however. I simply said, in an almost matter-of-fact way, that these cases seemed "interesting from a privacy perspective."

The NYU team took the suggestion and ran with it. In an amicus brief to the Fourth Circuit Court of Appeals, the team focused part of their argument around supplemental cases. They highlighted how the earlier positive decision on the spy plane's constitutionality from the court "relied heavily on the terms of the MOU. . . . Yet . . . critical aspects of the AIR program operate quite differently."[6] In other words, what the BPD said the plane *would* do and what it was *actually* doing were different. The finding was a key part of their final report, where they concluded, "The Supplemental Requests arguably exceed the terms of the MOU, and certainly exceed what people understood as the intended use [of the plane]."[7] They detailed how the program tracked people's mothers, watched people go shopping, and tracked folks who were nowhere near or even had anything to do with a crime scene. These empirical observations about differences between the stated and actual investigative process were noted in the court's decision to overturn its original ruling. The court changed its mind and called the spy plane unconstitutional.

Later, the NYU team thanked me in an email for the assistance. "We decided to keep people's names out of the report entirely," one of the authors wrote, "but I personally wanted to thank you for giving us some valuable context for our work." What I told the NYU team clearly, though perhaps indirectly, intervened in the course of events. Would the auditors have seen and understood the significance of "supplementals" had I not pointed them there? Perhaps. But there is no question that our conversation sped that process along.

As a qualitative social researcher, trained in the positivist environment of American academia, I have conflicting thoughts about this decision. Privacy advocates would probably say, yes, this was right. I simply gave the most accurate information to a team of researchers who were hired to get the most accurate information. Hard-core empiricists would probably say, no, this was wrong. By intervening here, I influenced the way the experiment played out. But it isn't as simple as even those two positions, in my mind. There is also an ethical objection here, especially knowing how the Fourth Circuit's ruling impacted the people who were experimented upon with the plane. The ruling was a key factor in throwing out several horrific murder cases, causing the evidence to become "fruit of the poisonous tree" and dashing the hopes of grieving family members. How should I think about my role in that chain of events? I leave it up to the reader to judge my actions here.

How Not to Get Subpoenaed

Given my insider access, it quickly became obvious that my field notes could get subpoenaed—either in the civil trial against the BPD or in one of the many criminal cases I witnessed. This raised a significant ethical dilemma: how could I protect the confidentiality of my research participants in the event that a judge asked for my field notes? In particular, I was concerned about exposing the names and actions of the many analysts who felt deep fear about going public as PSS workers. Recall that many Black analysts took extreme measures to prevent themselves from being labeled a "snitch" and suffering retributive violence. Ideally, if a judge asked me to hand over my field notes, I would be able to say no. I had promised all respondents, other than those who were already publicly identified with the program, that I would protect their confidentiality. I used pseudonyms throughout the field notes and the book. I changed some of the key details about the investigations I analyzed. Still, if you have insider knowledge, like a judge or attorney does, it would not be that hard to reconstruct people's identities by connecting the dots.

It's more than that, though. Giving my field notes over for inspection would not only betray the promises I made to participants, but also jeopardize the whole model of embedded, behind-the-scenes research like the kind I was doing. What tech company would ever again allow an ethnographer in like this, if they knew the promises of confidentiality were bogus? I wanted to protect respondents, but I also wanted to keep the pathway clear for future researchers.

I went to the literature for answers. It turns out there is a long line of ethnographers who have found themselves in this position.[8] It's not a pretty picture. To begin with, colleges and universities typically abandon qualitative researchers when they ask for legal backing against subpoena. Institutions usually say that they can help researchers provide "limited confidentiality" to participants. That means helping researchers protect confidentiality "to the extent that the law allows."[9] In other words, if you get subpoenaed, the institution will throw up its hands and offer no assistance.

When I approached my institution about this potential problem, first through the Institutional Review Board and then the general council, I was politely told that they would not defend me if I chose to withhold field notes. If a judge asked, I would need to comply. Defending myself would be seen by the college as a private legal matter. This is how it goes, I found out from the literature. Researchers have spent tens of thousands of dollars of their own money on legal representation defending themselves against a subpoena.[10] At least one researcher has even done jail time for refusing to comply with a judge's command.[11] While this literature gave me some cold comfort that I wasn't alone, it was also not very helpful. Since I was neither wealthy enough to pay for a lawyer nor principled enough to do jail time, what was I supposed to actually *do* to head off this problem? I needed practical advice.

Then, I discovered Ted and John. Ted Palys and John Lowman, Canadian criminologists, have dedicated their careers to the problem of subpoena and confidentiality in qualitative research. They have published dozens of articles and books on the topic. All ethnographers who work in sensitive field sites, especially around crime, ought to become familiar with their work. I reached out to Palys and Lowman. They immediately replied and offered to help me think through how to protect myself in the event of a subpoena. Here's what I found out.

The first thing they highlighted is that the legal teams at US and Canadian academic institutions base their inability to protect researchers on a narrow reading of a few unusual cases. The most common one is known as the Belfast Project case.[12] In the early 2000s, Boston College Library became involved in working with Irish journalists to record and preserve interviews with combatants from the conflict in Northern Ireland known as "the Troubles." The interviews were so sensitive, including descriptions of and confessions to murders, that the researchers used a random letter, for example, respondent *A*, to refer to their participants. When some of the interviews were finally made public, following the deaths of a few interviewees, Northern Ireland police were made aware of the project by a victim's family members. Police realized there was a trove of interviews germane to their investigations sitting in Boston. They subpoenaed the library for the original tapes. They were thus asking researchers to break confidentiality.

Boston College lawyers worked with the researchers to push back against the subpoena. After an extremely long, expensive, and acrimonious defense, the college lost. Researchers ended up having to comply with the subpoena, the pseudonymous interviewees were unmasked, and the tapes were used as evidence in criminal trials in Northern Ireland.

Why did the college lose the challenge? Palys and Lowman say it's not because the college promised protections of confidentiality that wouldn't actually stand up in court, it's because the Northern Irish court did things no US (or Canadian) court would do.[13] Palys, in an email communication with me, said the court went on a kind of "fishing expedition," rummaging through all the tapes without a clear reason until they found useful information. They also didn't attempt to find similar information through a nonconfidential source before going to the Belfast Project tapes. That would never fly in a US (or Canadian) court. In all likelihood, had the Belfast Project been adjudicated in a more traditional court (and had Boston College lawyers pushed back harder), the library researchers would have won.

Instead of the Belfast Project, then, Palys and Lowman suggest university legal teams look at more relevant cases where courts have operated more reliably, according to common law. One important case is *Cusumano v. Microsoft Corporation*. Here, two well-established qualitative researchers, Michael Cusumano and David Yoffie, were subpoenaed in an antitrust case for confidential interviews they conducted with the employees behind the 1990s "browser wars" between Netscape and Explorer.[14] Cusumano and Yoffie pushed back. They argued that the need to protect academic freedom as a public good outweighed the interests of Microsoft. Moreover, Microsoft could get the information they needed in other ways, without having to look through the researchers' confidential recordings and notes. In a way, Microsoft was asking the researchers to do their dirty work for them, when the company could just go out, on their own time and dime, and interview the same people the researchers did.

Cusumano and Yoffie won, even after Microsoft appealed to a federal court. Why? The court said that qualitative academic researchers ought to have the same level of protection as that given to journalists. The judges ruled,

> Journalists are the personification of free press, and to withhold such protection would invite a "chilling effect on speech," and thus destabilize the First Amendment. The same concerns suggest that courts ought to offer similar protection to academicians engaged in scholarly research. . . . As with reporters, a drying-up of sources would sharply curtail the information available to academic researchers and thus would restrict their output. Just as a journalist, stripped of sources, would write fewer, less incisive articles, an academician, stripped of sources, would be able to provide fewer, less cogent analyses.[15]

Rather than look at the unique Belfast Project case for guidance, which is actually an outlier, university legal teams should look here. A US federal court has held that academic freedom protects the public's right to know and protecting confidential research notes and recordings from subpoena is crucial to upholding that right. In my case, then, that means protecting the public's right to know what's going on with the spy plane, made possible by a promise of confidentiality to employees, outweighs a court's need to get information on a specific civil or criminal case.[16] There are many, many other cases, too, that have turned out favorably for researchers trying to protect their field notes from subpoena.[17] Despite what college lawyers often tell ethnographers, and what ethnographers themselves report in the literature, things are actually looking pretty good for research participant confidentiality.

Still, it really should never have to come to this. There are practical steps researchers can take, especially in the early phases, to protect themselves from ever having to make these kinds of protracted and expensive defenses.[18] First and foremost, if researchers are doing health-related or crime-related research, they can try to get a Certificate of Confidentiality from the National Institute of Health or a Privacy Certificate from the National Institute of Justice. Having this kind of document before you start means, as Palys told me, "you're home free." Unfortunately, certificates are almost always dependent on getting competitive funding from these institutions. If a researcher *can* get this funding, however, it will help bolster the claim that the researcher's work is for the public good. A breach of confidentiality, it can be claimed, would jeopardize work that is being funded by a public institution with public money.

If you don't have an official certificate and the financial backing of a big institute, Palys and Lowman suggest gathering data according to the principles of the Wigmore test—a Canadian legal test for balancing confidentiality and disclosure that is also followed in US courts.[19] Practically, this means doing the following.

ENSURE CONFIDENTIALITY IS MORE THAN WORDS ON A FORM

Throughout the process of talking with participants, confidentiality must be *explicitly* offered in words and *adhered* to in practice. Both sides need to know, before they talk, that everything uttered is in confidence, and researchers need to stick to that promise. It needs to be more than a form that someone signs. Researchers need to be able to *document* that they have not shared confidential information outside of the researcher-participant relationship. Do not talk to a partner, friend, or other people outside the study about confidential information.

Have a continuous chain of secure protection and encryption for field notes and recordings.

DOCUMENT THE NECESSITY OF CONFIDENTIALITY

In conversations with research participants, explicitly discuss the degree to which confidentiality is necessary for the researcher-participant relationship to hold. If respondents knew that the researcher could not provide confidentiality, would they refuse to talk? Would they withhold information or lie? Would the quality and reliability of the data be compromised? Researchers can ask about this in an interview and get it on the record. Documenting this information can help bolster claims that confidentiality is essential to upholding the public's right to the free flow of information. It can also help strengthen the claim that a breach of confidentiality would ruin access for future researchers.

MAKE TWO SETS OF FIELD NOTES

For researchers who write things down during participant observation, figure out a way to make the original notes difficult to access through the court's disclosure process. Practically, this can mean making digital copies of original field notes with redactions, de-identifications, or any other adjustments that need to be made. These less sensitive notes can be stored in an encrypted setting and used for analysis and final write-up. Then, send the original paper field notes, or digital field notes on a storage device, out of the country. *Do not tell anyone that the more sensitive notes exist.* This procedure helps reinforce the claim that there is nothing of interest for the court in the materials held by the researcher and that similar information could be gotten through a less sensitive source. But more importantly, even if the court did look at the materials, they won't be helpful because they don't contain any identifying information.

ORGANIZE TO CHANGE "LIMITED CONFIDENTIALITY" POLICIES

Policies that say that the university will help researchers protect participants' confidentiality "to the extent the law allows" are bad for academic freedom. They are designed to "cover" the university, not protect the free flow of information for the public good. Palys and Lowman call this "caveat emptor ethics," where the university gets to benefit from the prestige of the research and then walk away when things get tricky for the researcher.[20] Qualitative researchers on a given campus and across institutions can band together to push for better. Policies need to be designed to protect academic freedom first and foremost, not save the university time and money. Professional associations, such as the American Sociological Association, should help here by crafting model language for universities to follow. Better yet, they could develop their own legal team, which specializes in helping researchers navigate these issues, aiding individual ethnographers as well as Institutional Review Boards.[21]

In the end, I never did get subpoenaed. When all the spy plane cases were thrown out, I breathed a sigh of relief. Still, having followed the advice from Ted and John, I know that at least I was well prepared. I hope these reflections will help those who come after feel similarly prepared.

Notes

Chapter 1

1. All identifying details of the people and specific places related to criminal investigations in this book have been anonymized or altered in order to protect confidentiality. I use the real names of neighborhoods, however, in order to illustrate the general geographic locations of spy plane investigations. I only use the real names of people who have already become publicly identified with the AIR program in the media. For more details on the ethical framework that guided this research, see the appendix.

2. Stephen F. Henderson, "Fourth Amendment Time Machines (and What They Might Say about Police Body Cameras)," *Journal of Constitutional Law* 18, no. 3 (2016): 934.

3. J. Cavanaugh Simpson, "Prying Eyes: Military-Grade Surveillance Keeps Watch Over Baltimore and City Protests, but Catches Few Criminals," *Baltimore Magazine*, August 5, 2020, https://www.baltimoremagazine.com/section/community/surveillance-planes-watch-over-baltimore-but-catch-few-criminals/.

4. On the concept of race-class subjugated communities, see Joe Soss and Vesla Weaver, "Police Are Our Government: Politics, Political Science, and the Policing of Race-Class Subjugated Communities," *Annual Review of Political Science* 20 (2017): 565–91.

5. Benjamin Shestakofsky, *Behind the Startup: How Venture Capital Shapes Work, Innovation, and Inequality* (Oakland: University of California Press, 2024).

6. Fred Turner, *From Counterculture to Cyberculture: Stewart Brand, the Whole Earth Network, and the Rise of Digital Utopianism* (Chicago: University of Chicago Press, 2006).

7. For a discussion of the infallible state assumption, see Sarah Brayne, *Predict and Surveil: Data, Discretion, and the Future of Policing* (New York: Oxford University Press, 2020), 54.

8. Evgeny Morozov, *To Save Everything, Click Here: The Folly of Technological Solutionism* (New York: Public Affairs, 2013).

9. Sally Goldenberg and Joe Anuta, "'Big Brother Is Protecting You': Eric Adams Pledges Stronger Policing, More Technology in 2023," *Politico*, December 24, 2022, https://www.politico.com/news/2022/12/24/eric-adams-policing-technology-new-york-00075359.

10. Steven Zeitchik, "AI May Be Searching You for Guns the Next Time You Go Out in Public," *Washington Post*, May 25, 2022, https://www.washingtonpost.com/technology/2022/05/20/evolv-metal-detectors-gun-detection/.

11. Marcia Kramer, "In Wake of Brooklyn Mass Shooting, Mayor Adams Says NYC Has Been Exploring Technologies to Keep People Safe in Subway System," *CBS News*, April 13, 2022, https://www.cbsnews.com/newyork/news/in-wake-of-brooklyn-subway-shooting-mayor-adams-says-nyc-has-been-exploring-technologies-to-keep-people-safe-in-subway-system/.

12. Harvey Molotch, *Against Security: How We Go Wrong at Airports, Subways, and Other Sites of Ambiguous Danger* (Princeton, NJ: Princeton University Press, 2012).

13. "Statement on AI Risk: AI Experts and Public Figures Express Their Concern about AI Risk," Center for AI Safety, March 30, 2023, https://www.safe.ai/statement-on-ai-risk.

14. Michel Foucault, *Discipline and Punish: The Birth of the Prison*, trans. Alan Sheridan (New York: Pantheon Books, 1977).

15. Arthur Rizer, "Very Little Stands between the U.S. and a Technological Panopticon," *Slate*, November 19, 2020, https://slate.com/technology/2020/11/law-enforcement-facial-recognition-technology.html.

16. Lee Vinsel, "You're Doing It Wrong: Notes on Criticism and Technology Hype," *STS News* (blog), February 1, 2021, https://sts-news.medium.com/youre-doing-it-wrong-notes-on-criticism-and-technology-hype-18b08b4307e5.

17. For an example of criti-hype in academic scholarship, see Shoshana Zuboff, *The Age of Surveillance Capitalism: The Fight for a Human Future at the New Frontier of Power* (New York: Public Affairs, 2019).

18. "Gartner Hype Cycle: Explained," YouTube, 2022, technology video, 3:36, https://www.youtube.com/watch?v=jB1RDz9jajo.

19. Catherine O'Neil, *Weapons of Math Destruction: How Big Data Increases Inequality and Threatens Democracy* (New York: Crown Random House, 2016).

20. Scott O. Lilienfeld and Kristin Landfield, "Science and Pseudoscience in Law Enforcement: A User-Friendly Primer," *Criminal Justice and Behavior* 35, no. 10 (2008): 1215–30.

21. Jenna Burrell, "Artificial Intelligence and the Ever-Receding Horizon of the Future," *Tech Policy Press*, June 6, 2023, https://techpolicy.press/artificial-intelligence-and-the-ever-receding-horizon-of-the-future/.

22. "AI Hype Distracted Us from Real Problems, with Timnit Gebru," *Tech Won't Save Us* (podcast), January 18, 2024, https://techwontsave.us/episode/203_ai_hype_distracted_us_from_real_problems_w_timnit_gebru.

23. Ruha Benjamin, *Captivating Technology: Race, Carceral Technoscience, and Liberatory Imagination in Everyday Life* (Durham, NC: Duke University Press, 2019), 12.

24. A similar thread can be found in the debate over so-called artificial intelligence; for example, Arvind Narayanan and Sayash Kapoor, *AI Snake Oil: What Artificial Intelligence Can Do, What It Can't, and How to Tell the Difference* (Princeton, NJ: Princeton University Press, 2024); Frederike Kaltheuner, ed., *Fake AI* (Manchester, UK: Meatspace Press, 2021).

25. Virginia Eubanks, *Automating Inequality: How High-Tech Tools Profile, Police, and Punish the Poor* (New York: St. Martin's Press, 2017); Sarah Brayne, "Big Data Surveillance: The Case of Policing," *American Sociological Review* 82, no. 8 (2017): 977–1008; Ruha Benjamin, *Race after Technology: Abolitionist Tools for the New Jim Code* (New York: Polity, 2019); Brayne, *Predict and Surveil;* Sarah Brayne and Angele Christin, "Technologies of Crime Prediction: The Reception of Algorithms in Policing and Criminal Courts," *Social Problems*, 2020, https://doi.org/10.1093/socpro/spaa004; Sarah Esther Lageson, *Digital Punishment: Privacy, Stigma, and the Harms of Data-Driven Criminal Justice* (New York: Oxford University Press, 2020); Bryce Clayton Newell, *Police Visibility: Privacy, Surveillance, and the False Promise of Body-Worn Cameras* (Berkeley: University of California Press, 2021).

26. O'Neil, *Weapons of Math Destruction;* Joy Boulamwini and Timnit Gebru, "Gender Shades: Intersectional Accuracy Disparities in Commercial Gender Classification," *Proceedings of Machine Learning Research* 81 (2018): 1–15; Safia Umoja Noble, *Algorithms of Oppression: How Search Engines Reinforce Racism* (New York: NYU Press, 2018); Meredith Broussard, *Artificial Unintelligence: How Computers Misunderstand the World* (Cambridge, MA: MIT Press, 2019); Emily Bender et al., "On the Dangers of Stochastic Parrots: Can Language Models Be Too Big?," in *FAccT '21: Proceedings of the 2021 ACM Conference on Fairness, Accountability, and Transparency*, March 2021, 610–23, https://doi.org/10.1145/3442188.3445922;

Inioluwa Deborah Raji et al., "The Fallacy of AI Functionality," in *FAccT '22: Proceedings of the 2022 ACM Conference on Fairness, Accountability, and Transparency*, June 2022, 959-72, https://doi.org/10.1145/3531146.3533158; Meredith Broussard, *More than a Glitch: Confronting Race, Gender, and Ability Bias in Tech* (Cambridge, MA: MIT Press, 2023); Joy Boulamwini, *Unmasking AI: My Mission to Protect What Is Human in a World of Machines* (New York: Random House, 2023); Tamara Kneese, *Death Glitch: How Techno-Solutionism Fails Us in This Life and Beyond* (New Haven, CT: Yale University Press, 2023).

27. Notable organizations include the Stop LAPD Spying Coalition in Los Angeles, Lucy Parsons Labs and the Invisible Institute in Chicago, the Surveillance Technology Oversight Program in New York City, Open Justice Baltimore, and the Center on Privacy and Technology at Georgetown Law School, among many others.

28. Brayne, *Predict and Surveil*, 4.

29. For important exceptions, see Eubanks, *Automating Inequality*; Brayne, *Predict and Surveil*; Brayne and Christin, "Technologies of Crime Prediction"; Lageson, *Digital Punishment*; Newell, *Police Visibility*.

30. Arthur Holland Michel, *Eyes in the Sky: The Secret Rise of Gorgon Stare and How It Will Watch Us All* (New York: Houghton Mifflin Harcourt, 2019); Monte Reel, "Secret Cameras Record Baltimore's Every Move from Above," *Bloomberg Businessweek*, August 23, 2016, https://www.bloomberg.com/features/2016-baltimore-secret-surveillance/.

31. Manoush Zomorodi and Alex Goldmark, "Eye in the Sky," *Radiolab* (podcast), WNYC Studios, June 18, 2015, https://www.wnycstudios.org/story/eye-sky.

32. This research has been previously published. See Benjamin H. Snyder, "'Big Brother's Bigger Brother': The Visual Politics of (Counter) Surveillance in Baltimore," *Sociological Forum* 35, no. 4 (2020): 1315-36.

33. Barry Friedman et al., *Civil Rights and Civil Liberties Audit of Baltimore's Aerial Investigation Research (AIR) Program* (New York: Policing Project at NYU Law, 2020), https://www.policingproject.org/s/AIR-Program-Audit-Report-vFINAL-reduced.pdf; Ann P. Cotton et al., *Baltimore Aerial Investigation Research Project: Findings from the Early Launch Community Survey* (Baltimore: Schaefer Center for Public Policy, University of Baltimore, June 2020), https://68i.ab1.myftpupload.com/wp-content/uploads/2020/12/AIRCommunitySurveyReport-SchaeferCenter-FINAL.pdf; Andrew R. Morral et al., *Evaluating Baltimore/s Aerial Investigation Research Pilot Program: Interim Report* (Santa Monica, CA: RAND, 2021), https://www.rand.org/pubs/research_reports/RRA1131-2.html.

34. In this way, I purposefully did not follow more recent trends in ethnography to engage in "observant participation," where the researcher seeks to reduce

the distance between the observer and the observed. Josh Seim, "Participant Observation, Observant Participation, and Hybrid Ethnography," *Sociological Methods and Research*, 2021, https://doi.org/10.1177/0049124120986209.

35. Colin Jerolmack and Shamus Khan, "Talk Is Cheap: Ethnography and the Attitudinal Fallacy," *Sociological Methods and Research* 43, no. 2 (2014): 178–209.

36. Douglas Freitag, Terry Wohlers, and Therese Philippi, *Rapid Prototyping: State of the Art* (Chicago: Manufacturing Technology Information Analysis Center, October 23, 2003), https://apps.dtic.mil/sti/citations/ADA435248.

37. Shestakofsky, *Behind the Startup*.

38. Other names I've seen are wide area surveillance (WAS) and wide area persistent surveillance (WAPS).

39. Michel, *Eyes in the Sky*.

40. Derek Gregory, "From a View to a Kill: Drones and Late Modern War," *Theory, Culture and Society* 28, no. 7–8 (2011): 188–215.

41. Paul Virilio, *War and Cinema: The Logistics of Perception* (London: Verso, 2009).

42. Michel, *Eyes in the Sky*.

43. Reel, "Secret Cameras Record Baltimore's Every Move from Above."

44. The company's brief contracts in Juárez, Dayton, Compton, and Philadelphia, which stretched between 2009 and 2015, have been described nicely elsewhere. Conor Friedersdorf, "Eyes Over Compton: How Police Spied on a Whole City," *The Atlantic*, April 21, 2014, https://www.theatlantic.com/national/archive/2014/04/sheriffs-deputy-compares-drone-surveillance-of-compton-to-big-brother/360954/; Michel, *Eyes in the Sky*; Reel, "Secret Cameras Record Baltimore's Every Move from Above."

45. Matthew Guariglia and Dave Maass, "How Police Fund Surveillance Technology Is Part of the Problem," Electronic Frontier Foundation, September 23, 2020, https://www.eff.org/deeplinks/2020/09/how-police-fund-surveillance-technology-part-problem.

46. Lily Hay Newman, "How Baltimore Became America's Laboratory for Spy Tech," *Wired*, September 4, 2016, https://www.wired.com/2016/09/baltimore-became-americas-testbed-surveillance-tech/.

47. Kevin Rector, "Council Plans Hearing on Police Surveillance," *Baltimore Sun*, August 26, 2016, sec. 1A; Doug Donovan, "Private Fund, Public Mistrust," *Baltimore Sun*, August 28, 2016, sec. 1A.

48. Kevin Rector and Justin George, "Firm Urged City Police to Go Public on Surveillance," *Baltimore Sun*, October 21, 2016, sec. 1A.

49. Reel, "Secret Cameras Record Baltimore's Every Move from Above."

50. McNutt told me that when Reel approached him to participate in the article, giving the reporter inside access to the operations center, he consented out of a commitment to "total transparency" (as he would later do with me), even though he knew the BPD would not approve.

51. Editorial Board, "Big Data Is Watching," *Baltimore Sun*, October 24, 2016, sec. 10A.

52. National Policing Institute, *A Review of the Baltimore Police Department's Use of Persistent Surveillance (Baltimore Community Support Program)* (Arlington, VA: National Policing Institute, January 30, 2017).

53. Talia Richman, "Baltimore Aerial Surveillance Agreement: $3.7 Million Price Tag, Privacy Protections, Evaluation Plan," *Baltimore Sun*, March 25, 2020, https://www.baltimoresun.com/politics/bs-md-pol-aerial-surveillance-agreement-boe-20200324-lvpjbsvqs5catntaeva2532a2a-story.html.

Chapter 2

1. Daniel S. Nagin, "Deterrence in the Twenty-First Century," *Crime and Justice* 42, no. 1 (2013): 199–263.

2. Weihua Li and Jamiles Lartey, "As Murders Spiked, Police Solved about Half in 2020," *Marshall Project* (blog), January 12, 2022, https://www.themarshallproject.org/2022/01/12/as-murders-spiked-police-solved-about-half-in-2020.

3. Jeremy Kahn, "The Story of a Snitch," *The Atlantic*, April 2007, https://www.theatlantic.com/magazine/archive/2007/04/the-story-of-a-snitch/305703/.

4. U.S. Department of Justice, *Investigation of the Baltimore City Police Department* (Washington, DC: U.S. Department of Justice, Civil Rights Division, August 10, 2016).

5. John Browning, "#Snitches Get Stitches: Witness Intimidation in the Age of Facebook and Twitter," *Pace Law Review* 35, no. 1 (2014): 192–214.

6. Justin Fenton, "Instagram Shut Accounts Used to Intimidate Witnesses in Baltimore; Hours Later New One Popped Up," *Baltimore Sun*, December 2, 2020, https://www.baltimoresun.com/2020/12/02/instagram-shut-accounts-used-to-intimidate-witnesses-in-baltimore-hours-later-new-one-popped-up/.

7. One detective I interviewed noted that, these days, much of the evidence detectives obtain through witness statements comes from arrests on other cases. When a person is picked up on drug or weapons possession, for example, they will provide information on, say, a homicide case as leverage for leniency. Even the witnessing that does come from Baltimore citizens, then, is often given under coercive conditions, not out of a sense of trust or obligation. The detective be-

moaned that, as the BPD has been forced to reform itself, officers have been instructed to carry out fewer arrests for low-level offenses (especially marijuana possession). Thus they have had fewer opportunities to gather information from this kind of coerced, witnessing as leverage.

8. Though PSS also had access to the live feed of the CitiWatch system, they were instructed never to use it.

9. Christopher Slobogin, *Virtual Searches: Regulating the Covert World of Technological Policing* (Oakland: University of California Press, 2022).

10. Jay Stanley and Ross McNutt, "CFP 2014: Persistent Aerial Surveillance," YouTube, June 16, 2014, slide presentation, 53:22, https://youtu.be/gYhjktrmPuA.

11. According to the study, on simple comparison between cases closed with the spy plane and cases closed without, the spy plane helped. Spy plane–involved cases performed about 8 percent better (statistically significant). But when RAND broke the results down into a more apples-to-apples comparison, using a subset of non–spy plane cases that were more similar to the kinds of cases PSS worked, spy plane cases were not closed at a statistically significantly higher rate. Andrew R. Morral et al., *Preliminary Findings from the Aerial Investigations Research Pilot Program* (Santa Monica, CA: RAND, 2021), 3–5.

12. Sarah Brayne, *Predict and Surveil: Data, Discretion, and the Future of Policing* (New York: Oxford University Press, 2020), 6.

13. Paul M. Leonardi, "Materiality, Sociomateriality, and Socio-Technical Systems: What Do These Terms Mean? How Are They Different? Do We Need Them?," in *Materiality and Organizing: Social Interaction in a Technological World*, ed. Paul M. Leonardi, Bonnie A. Nardi, and Jannis Kallinikos (New York: Oxford University Press, 2012), 24–48.

Chapter 3

1. Tyler Waldman, "Harrison: Surveillance Plane to Return for Trial Program Next Year," *WBAL News*, December 20, 2019, https://www.wbal.com/article/427127/124/harrison-surveillance-plane-to-return-for-trial-program-next-year.

2. Recordings of the public sessions were accessed here: *Aerial Investigation Research Pilot Program* (Baltimore, 2020), accessed March 24, 2022, https://www.facebook.com/watch/live/?ref=watch_permalink&v=3400646286628872.

3. Pseudoscience is common in law enforcement. The BPD's use of the spy plane exhibited several of the classic warning signs of a pseudoscientific technique, including overuse of ad hoc maneuvers to explain insignificant impact,

evasion of peer review, persistence of false claims that were resistant to self-correction, extravagant claims, lack of documentation of subjective judgments that led to a decision, and the absence of safeguards against confirmation bias. See Scott O. Lilienfeld and Kristin Landfield, "Science and Pseudoscience in Law Enforcement: A User-Friendly Primer," *Criminal Justice and Behavior* 35, no. 10 (2008): 1216.

4. Privacy risks, which I discuss at length in chapter 6, were also downplayed.

5. Conversations about the program took place across social media during March 2020. Reddit featured a particularly nuanced debate about the legality and efficacy of the program. One user noted, "I am curious if the plane's effectiveness will be accurately determined. The current covid situation is by no means normal, so it could make the surveillance appear more or less effective than expected, depending on the data points used." Another user replied, "The measured effectiveness is a huge part of it. I'm hoping that it will prove ineffective and we can move on from this idea. It just doesn't pass the sniff test as far as its creepiness."

6. McNutt expressed many of these same concerns to me when he saw Harrison's presentation, saying that he wished the commissioner had let him show a full investigation. Police refused and told McNutt they would handle the public rollout alone.

7. Eduardo Bonilla-Silva, *Racism without Racists: Color-Blind Racism and the Persistence of Racial Inequality in America*, 6th ed. (New York: Rowman and Littlefield, 2021).

8. As I explore more in the next chapter, much like other camera systems in the city, such as CitiWatch, the spy plane was invested in the dual tasks of protecting property in White Baltimore and punishing gun violence in Black Baltimore.

9. Lilienfeld and Landfield, "Science and Pseudoscience in Law Enforcement," 1216.

10. There is a lot of missing information from this case, which I could not gather from interviews and internal documents. All the tracking that resulted in a determination that a car or person was not involved (called DNI tracks) were not recorded in the briefing document or in internal notes. That information can only be obtained from the raw footage, which police would not give me access to after the pilot program finished. Therefore, I have had to reconstruct some of these events. I have invented some of the information in order to convey the general contours of the investigation in a clear narrative. I have flagged where information is real and where it is invented.

11. There are two exceptions to Sandtown's CCTV desert. One is a large public housing complex, Gilmore Homes, on the northern edge of the neighborhood,

where a tight clutch of cameras captures not just the public streets surrounding the complex but also the semiprivate recreation areas between buildings. The second exception is two cameras on the outside of the Western District police headquarters.

12. I was able to confirm the exact location of this apartment complex with Detective Karim, so the easternmost point of the dashed black line in map 2 is accurate.

13. Stopping a person for looking suspicious in a high-crime area is part of the discretionary power given to police by current Fourth Amendment jurisprudence. Illinois v. Wardlow, No. 528 U.S. 119 (January 12, 2000).

14. Detective Karim disagrees with this interpretation. He admits that "harass" was not the best choice of words during the interview. "I also use that word to describe making fun of [a person] good-naturedly," he notes. His impression of the encounter was that "no one was upset by our presence, as far as I know." Without interviewing everyone involved, however, I cannot know for sure.

15. I talked to many policing experts outside Baltimore who found this practice of making arrests based on "visual similarity" in two pieces of camera footage to be unconvincing. As one expert told me, "There are lots of people who wear similar sorts of clothing or have similar builds out on the street every day." The notion that two figures with similarly colored clothing, of "medium build," and with "dark skin," for example, are definitively the same person is not particularly convincing to investigators outside Baltimore.

16. The cars that Sarah mistakenly tracked, for example, she remembers being relabeled as DNI (determined not involved) in the raw tracking footage, as was standard procedure. Without the raw footage, I cannot confirm this. Still, the tracks of these DNI vehicles are not included in the final briefing document, which is what goes into evidence with the case and becomes part of the disclosure process during pretrial preparations.

17. Catherine O'Neil, *Weapons of Math Destruction: How Big Data Increases Inequality and Threatens Democracy* (New York: Crown Random House, 2016), 20–21.

18. Benjamin Shestakofsky, *Behind the Startup: How Venture Capital Shapes Work, Innovation, and Inequality* (Oakland: University of California Press, 2024).

19. This tendency to not see the "worst" coming because of a culture of positivity, what Cerulo calls "positive asymmetry," is typical of many organizations. See Karen A. Cerulo, *Never Saw It Coming: Cultural Challenges to Envisioning the Worst* (Chicago: University of Chicago Press, 2006).

20. Lilienfeld and Landfield, "Science and Pseudoscience in Law Enforcement," 1220.

Chapter 4

1. Lawrence T. Brown, *The Black Butterfly: The Harmful Politics of Race and Space in America* (Baltimore: Johns Hopkins University Press, 2021).

2. Lawrence Brown, "Two Baltimores: The White L vs. the Black Butterfly," *Baltimore Sun*, June 28, 2016, https://www.citypaper.com/bcpnews-two-baltimores-the-white-l-vs-the-black-butterfly-20160628-htmlstory.html.

3. Brett Theodos, Eric Hangen, and Brady Meixell, "The Black Butterfly: Racial Segregation and Investment Patterns in Baltimore," Urban Institute, February 5, 2019, https://apps.urban.org/features/baltimore-investment-flows/.

4. Garrett Power, "Apartheid Baltimore Style: The Residential Segregation Ordinances of 1910–1913," *Maryland Law Review* 42, no. 2 (1983): 289–329.

5. Power, "Apartheid Baltimore Style," 289–329; George Lipsitz, *How Racism Takes Place* (Philadelphia: Temple University Press, 2011); Richard Rothstein, *The Color of Law: A Forgotten History of How Our Government Segregated America* (New York: W. W. Norton, 2017).

6. Brown, "Two Baltimores."

7. Joe Soss and Vesla Weaver, "Police Are Our Government: Politics, Political Science, and the Policing of Race-Class Subjugated Communities," *Annual Review of Political Science* 20 (2017): 565–91.

8. Isaac Ariail Reed, *Interpretation and Social Knowledge: On the Use of Theory in the Human Sciences* (Chicago: University of Chicago Press, 2011), 11.

9. Patrick Sharkey, *Uneasy Peace: The Great Crime Decline, the Renewal of City Life, and the Next War on Violence* (New York: W. W. Norton, 2018); Patrick Sharkey and Marsteller Alisabeth, "Neighborhood Inequality and Violence in Chicago, 1965–2020," *University of Chicago Law Review* 89, no. 2 (2022): 3.

10. Lipsitz, *How Racism Takes Place*, 28.

11. Power, "Apartheid Baltimore Style," 289–329.

12. Zora Neale Hurston, "The 'Pet Negro' System," *American Mercury* 56 (1943): 593–600.

13. Monica C. Bell, "Anti-Segregation Policing," *New York University Law Review* 95 (2020): 690.

14. Adam Malka, *The Men of Mobtown: Policing Baltimore in the Age of Slavery and Emancipation* (Chapel Hill: University of North Carolina Press, 2018), 62–68.

15. Jennie K. Williams, "Trouble the Water: The Baltimore to New Orleans Coastwise Slave Trade, 1820–1860," *Slavery and Abolition* 41, no. 2 (2020): 275–303.

16. Malka, *Men of Mobtown*, 64–68.

17. Malka, 67.

18. The main recipients of these newly professionalized, powerful policing jobs were White, working-class men, who were wooed by politicians on a platform of creating new jobs for "native born Whites" who were threatened by immigrant labor; Malka, *Men of Mobtown*, 63. See also Khalil Gibran Muhammad, *The Condemnation of Blackness: Race, Crime, and the Making of Modern Urban America*, 2nd ed. (Cambridge, MA: Harvard University Press, 2019); Matthew Guariglia, *Police and the Empire City: Race and the Origins of Modern Policing in New York* (Chapel Hill, NC: Duke University Press, 2023).

19. Department of Homeland Security, *Public Workshop CCTV: Developing Privacy Best Practices* (Arlington, VA: Department of Homeland Security, December 17, 2007), 7, https://www.dhs.gov/sites/default/files/publications/privacy_rpt_cctv_2007.pdf.

20. "Around the Region," *Baltimore Sun*, April 25, 1990, sec. 37.

21. Department of Homeland Security, *Public Workshop CCTV*, 8.

22. Bill Glauber, "TV Cameras Keep Watch over British Streets," *Baltimore Sun*, January 29, 1996, sec. 89.

23. Department of Homeland Security, *Public Workshop CCTV*, 8.

24. Quoted in Department of Homeland Security, 8.

25. Peter Hermann, "Cameras to Watch Streets of City," *Baltimore Sun*, January 20, 1996, sec. 1A.

26. Department of Homeland Security, *Public Workshop CCTV*, 8.

27. Cara Nusbaum, "Police, Downtown Group Credit Video Cameras with Helping to Reduce Crime," *Baltimore Sun*, March 13, 2001, sec. 17.

28. Associated Press, "Methadone Price Rises, Forcing Cut in Aid to Addicts," *Baltimore Sun*, November 9, 1990, sec. 19.

29. Hermann, "Cameras to Watch Streets of City," 5A.

30. Department of Homeland Security, *Public Workshop CCTV*, 8.

31. Department of Homeland Security, 8.

32. Michael R. Bromwich et al., *Anatomy of the Gun Trace Task Force Scandal: Its Origins, Causes, and Consequences* (Washington, DC: Steptoe, 2022), 19.

33. Department of Homeland Security, *Public Workshop CCTV*, 12.

34. Department of Homeland Security, 9.

35. Editorial Board, "No Substitute for Police Officers," *Baltimore Sun*, April 26, 1996, sec. 18.

36. Editorial Board, sec. 18.

37. La Quinta Dixon, "Greektown Gets Security Cameras in Spreading Effort to Reduce Crime," *Baltimore Sun*, July 29, 1999, sec. 4B.

38. On the history of US immigrants becoming White, see David R. Roediger, *Working Toward Whiteness: How America's Immigrants Became White* (New York: Basic Books, 2006).

39. Dixon, "Greektown Gets Security Cameras," sec. 4B.

40. Ana Muñiz, *Borderland Circuitry: Immigration Surveillance in the United States and Beyond* (Oakland: University of California Press, 2022); Melissa Villa-Nicholas, *Data Borders: How Silicon Valley Is Building an Industry around Immigrants* (Oakland: University of California Press, 2023).

41. Nancy G. La Vigne et al., *Evaluating the Use of Public Surveillance Cameras for Crime Control and Prevention* (Washington, DC: Urban Institute, September 2011), 24.

42. La Vigne et al., 24.

43. Gus S. Sentementes and John Fritze, "3 Hunted in Attack on Reporter," *Baltimore Sun*, March 1, 2006, sec. 1B.

44. Lori Barrett, "Baltimore Will Begin Phasing Out the Ugly and Largely Ineffective Blue-Light Cameras," *Baltimore Sun*, May 19, 2008, sec. 4T.

45. Matthew Guariglia and Dave Maass, "How Police Fund Surveillance Technology Is Part of the Problem," Electronic Frontier Foundation, September 23, 2020, https://www.eff.org/deeplinks/2020/09/how-police-fund-surveillance-technology-part-problem; David Lyon, "9/11, Synopticon, and Scopophilia: Watching and Being Watched," in *The New Politics of Surveillance and Visibility*, ed. Kevin D. Haggerty and Richard V. Ericson (Toronto: University of Toronto Press, 2006), 35–54.

46. Karen Hosler, "Maryland Likely to Receive $56 Million from Congress," *Baltimore Sun*, December 5, 2001, sec. 8A; Doug Donovan, "24-Hour Camera Surveillance in City Is Part of Bigger Plan," *Baltimore Sun*, June 10, 2004, sec. 1A.

47. Once again, London was a source of inspiration. Traveling to the United Kingdom in 2004, Mayor O'Malley and Baltimore police executives toured London's massive CCTV operations to see how a live-monitored system works. Donovan, "24-Hour Camera Surveillance," sec. 10A.

48. Doug Donovan, "Camera Network to Make Its Debut," *Baltimore Sun*, May 9, 2005, sec. 1B.

49. Donovan, "24-Hour Camera Surveillance," sec. 10A.

50. Editorial Board, "Watch Out," *Baltimore Sun*, June 17, 2004, sec. 18A.

51. Doug Donovan, "Camera System Expands in City," *Baltimore Sun*, December 2, 2004, sec. 1B.

52. Donovan, sec. 9B.

53. The total cost of the system is much larger. This figure comes from the Urban Institute's report, which estimates the "start-up cost" of the system in the year 2008 to be approximately $5.5 million and the total cost including labor and maintenance to be $8.1 million. Since 2008, many more cameras have been installed and the manpower of the system has increased. La Vigne et al., *Evaluating the Use of Public Surveillance Cameras*, 48.

54. Donovan, "Camera Network to Make Its Debut," sec. 4B. (Emphasis added.)

55. La Vigne et al., *Evaluating the Use of Public Surveillance Cameras*, 31, 36, 44.

56. La Vigne et al., 40.

57. See, for example, Kevin Rector, "Two Homes on Collapsed E. 26th St. Broken Into," *Baltimore Sun*, May 13, 2014, sec. 2A.

58. Editorial Board, "Not Big Brother," *Baltimore Sun*, September 21, 2011, sec. 18A.

59. Sarah Brayne, *Predict and Surveil: Data, Discretion, and the Future of Policing* (New York: Oxford University Press, 2020); Antony Loewenstein, *The Palestine Laboratory: How Israel Exports the Technology of Occupation Around the World* (New York: Verso, 2023).

60. Lily Hay Newman, "How Baltimore Became America's Laboratory for Spy Tech," *Wired*, September 4, 2016, https://www.wired.com/2016/09/baltimore-became-americas-testbed-surveillance-tech/.

61. Peter Hermann, "Police All Ears with Shot-Detection Program," *Baltimore Sun*, December 28, 2008, sec. 6A.

62. Justin Fenton, "System to Listen for Gunfire at Hopkins," *Baltimore Sun*, November 18, 2008, sec. 3A.

63. Michael Litch, *Draft Technical Report for SECURES Demonstration in Hampton and Newport News, Virginia* (Washington, DC: U.S. Department of Justice, Office of Justice Programs, January 2011), 54, https://www.ojp.gov/library/publications/draft-technical-report-secures-demonstration-hampton-and-newport-news-virginia.

64. Peter Hermann, "Jury Still Out on Use of Gunfire Detectors near Johns Hopkins," *Baltimore Sun*, December 21, 2008, sec. 6A.

65. Justin Fenton, "Police Test Gunshot-Detector," *Baltimore Sun*, October 21, 2009, sec. 8A.

66. Justin George and Yvonne Wenger, "Baltimore Police to Install Gunshot Detection System on East, West Sides," *Baltimore Sun*, February 13, 2014, sec. 2A.

67. Center for Investigative Reporting, "Is ShotSpotter Worth the Expense?," *Reveal* (podcast), accessed June 12, 2023, https://on.soundcloud.com/wXsVs.

68. Tim Prudente and Sarah Meehan, "Gunshot Detection System Unveiled," *Baltimore Sun*, June 2, 2018, sec. 2A.

69. Joseph M. Ferguson and Deborah Witzburg, *The Chicago Police Department's Use of ShotSpotter Technology* (Chicago: Office of Inspector General, August 2021).

70. Jay Stanley, "Four Problems with the ShotSpotter Gunshot Detection System," ACLU News & Commentary, August 24, 2021, https://www.aclu.org/news/privacy-technology/four-problems-with-the-shotspotter-gunshot-detection-system.

71. Jamie Kalven, "Chicago Awaits Video of Police Killing of 13-Year-Old Boy," *The Intercept*, April 13, 2021, https://theintercept.com/2021/04/13/chicago-police-killing-boy-adam-toledo-shotspotter/.

72. In 2024, just as this book was about to go to press, the locations of all ShotSpotter sensors in the United States, including in Baltimore, were leaked by a disgruntled employee. This is the only way that the geographical distribution of the technology has been made public. Dhruv Mehrotra and Joey Scott, "Here Are the Secret Locations of ShotSpotter Gunfire Sensors," *Wired*, February 22, 2024, https://www.wired.com/story/shotspotter-secret-sensor-locations-leak/.

73. Brayne, *Predict and Surveil*.

74. Harris's stingray device is shrouded in secrecy. The best estimate of a price I could find is "the low six figures." Sam Biddle, "Long-Secret Stingray Manuals Detail How Police Can Spy on Phones," *The Intercept*, September 12, 2016, https://theintercept.com/2016/09/12/long-secret-stingray-manuals-detail-how-police-can-spy-on-phones/.

75. Human Rights Watch, *Dark Side: Secret Origins of Evidence in US Criminal Cases* (New York: Human Rights Watch, 2018).

76. Justin Fenton, "Police Track Cellphones Secretly," *Baltimore Sun*, April 9, 2015, sec. 1A.

77. Justin Fenton, "Guilty Pleas Short-Circuit Debate over Cell Tracking," *Baltimore Sun*, January 8, 2015, sec. 1A.

78. Jack Gillum, "Baltimore FBI Agreement," Associated Press, July 13, 2011, https://www.documentcloud.org/documents/1809046-.

79. Fenton, "Police Track Cellphones Secretly," sec. 1A.

80. A log of cell phone "captures," obtained through a *USA Today* investigative report, can be found here: Brad Heath, "Cell Site Data Request 060815 Bds," *USA Today*, August 23, 2015, https://www.documentcloud.org/documents/2287407-cell-site-data-request-060815-bds-2.html.

81. Brad Heath, "Police Secretly Track Cellphones to Solve Routine Crimes," *USA Today*, August 23, 2015, https://www.usatoday.com/story/news/2015/08/23/baltimore-police-stingray-cell-surveillance/31994181/.

82. Jessica Anderson, "A New Challenge to Phone Tracking," *Baltimore Sun*, August 29, 2015, sec. 1A.

83. Laura Moy, *Complaint for Relief against Unauthorized Radio Operation and Willful Interference with Cellular Communications: Petition for an Enforcement Advisory on Use of Cell Site Simulators by State and Local Government Agencies* (Washington, DC: Georgetown University Law Center, August 16, 2016), https://www.documentcloud.org/documents/3013988-CS-Simulators-Complaint-FINAL.html#document/p4/a314050.

84. Prince Jones v. United States, No. 15-CF-322 (DC Ct. App. September 21, 2017).

85. Caroline Haskins, "There Are No Laws Restricting 'Stingray' Use; This New Bill Would Help," *Buzzfeed*, June 17, 2021, https://www.buzzfeednews.com/article/carolinehaskins1/new-law-restrict-stingray-surveillance-use.

86. Moy, *Complaint for Relief against Unauthorized Radio Operation*, 24.

87. Scott O. Lilienfeld and Kristin Landfield, "Science and Pseudoscience in Law Enforcement: A User-Friendly Primer," *Criminal Justice and Behavior* 35, no. 10 (2008): 1215–30.

88. Bell, "Anti-Segregation Policing," 690.

89. Technological border walls are common in the United States as a less visible technique for perpetuating political and social divides, for example, on the border between the United States and Mexico. See Muñiz, *Borderland Circuitry*; Villa-Nicholas, *Data Borders*.

90. Ed Vogel and Fletcher Nickerson, "Surveillance Technologies Don't Create Safety; They Intensify State Violence," *Truthout*, May 21, 2023, https://truthout.org/articles/surveillance-technologies-dont-create-safety-they-intensify-state-violence/.

Chapter 5

1. Baynard Woods and Brandon Soderberg, *I Got a Monster: The Rise and Fall of America's Most Corrupt Police Squad* (New York: St. Martin's Press, 2020); Justin Fenton, *We Own This City: A True Story of Crime, Cops, and Corruption* (New York: Random House, 2021).

2. Brandon Soderberg, "Baltimore Defense Attorneys Claim Surveillance Plane Footage Contradicts Law Enforcement Account of Police Shooting," *The*

Appeal, February 13, 2020, https://theappeal.org/baltimore-police-shooting-new-motion/.

3. Emphasis added.

4. Michael R. Bromwich et al., *Anatomy of the Gun Trace Task Force Scandal: Its Origins, Causes, and Consequences* (Washington, DC: Steptoe, 2022).

5. Soderberg, "Baltimore Defense Attorneys."

6. Justin Fenton, "Lawsuit Settlement Likely $8M," *Baltimore Sun*, November 14, 2020, sec. 1A.

7. Steve Mann, Jason Nolan, and Barry Wellman, "Sousveillance: Inventing and Using Wearable Computing Devices for Data Collection in Surveillance Environments," *Surveillance and Society* 1, no. 3 (2003): 331–55.

8. Kevin D. Haggerty and Ajay Sandhu, "The Police Crisis of Visibility," *IEEE Technology and Society Magazine*, 33, no. 2 (2014): 9–12.

9. Though I was not present for this meeting, PSS gave me access to a video recording. For more details and a deeper analysis of PSS's outreach efforts, see Benjamin H. Snyder, "'Big Brother's Bigger Brother': The Visual Politics of (Counter) Surveillance in Baltimore," *Sociological Forum* 35, no. 4 (2020): 1315–36.

10. Elijah Anderson, *Code of the Street: Decency, Violence, and the Moral Life of the Inner City* (New York: W. W. Norton, 1999).

11. U.S. Department of Justice, *Investigation of the Baltimore City Police Department* (Washington, DC: U.S. Department of Justice, Civil Rights Division, August 10, 2016).

12. Kevin Rector, "Baltimore Police Officer Found Guilty of Fabricating Evidence in Case Where His Own Body Camera Captured the Act," *Baltimore Sun*, November 9, 2018, https://www.baltimoresun.com/2018/11/09/baltimore-police-officer-found-guilty-of-fabricating-evidence-in-case-where-his-own-body-camera-captured-the-act/.

13. Simmons's comments resemble the "viewer society," described by Thomas Mathiesen. Thomas Mathiesen, "The Viewer Society: Michel Foucault's 'Panopticon' Revisited," *Theoretical Criminology* 1, no. 2 (1997): 215–34.

14. Anderson, *Code of the Street*.

15. Thomas Ward Frampton, "The Dangerous Few: Taking Seriously Prison Abolition and Its Skeptics," *Harvard Law Review* 135, no. 8 (2022): 2013–52.

16. U.S. Department of Justice, *Investigation of the Baltimore City Police Department*.

17. James Forman Jr., *Locking Up Our Own: Crime and Punishment in Black America* (New York: Farrar, Straus, and Giroux, 2017).

18. Lawrence Grandpre, "Who Speaks for the Community? Rejecting a False Choice between Liberty and Security," *Leaders of a Beautiful Struggle* (blog), June 5, 2020, https://lbsbaltimore.com/who-speaks-for-community-rejecting-a-false-choice-between-liberty-and-security/.

19. Some of the more vocal Black-led activist groups that came out against the program were Leaders of a Beautiful Struggle, Cease Fire 365, and Baltimore Bloc, as well as several "ally" groups, such as Showing Up for Racial Justice Baltimore and Open Justice Baltimore.

20. Ann P. Cotton et al., *Baltimore Aerial Investigation Research Project: Findings from the Early Launch Community Survey* (Baltimore: Schaefer Center for Public Policy, University of Baltimore, June 2020), https://68i.ab1.myftpupload.com/wp-content/uploads/2020/12/AIRCommunitySurveyReport-Schaefer-Center-FINAL.pdf.

21. Brittany Arsiniega and Matthew Guariglia, "Police as Supercitizens," *Social Justice* 48, no. 4 (2021): 33–58.

22. Geoffrey P. Alpert, Jeffrey J. Noble, and Jeff Rojek, "Solidarity and the Code of Silence," in *Critical Issues in Policing: Contemporary Readings*, ed. Roger G. Dunham, Geoffrey P. Alpert, and Kyle D. McLean (Long Grove, IL: Waveland Press, 2015), 106–21; David Weisburd et al., *Abuse of Police Authority: A National Study of Police Officers* (Washington, DC: Police Foundation, 2001).

23. Bryce Clayton Newell, "Context, Visibility, and Control: Police Work and the Contested Objectivity of Bystander Video," *New Media and Society* 21, no. 1 (2019): 60–76.

24. Kelly Gates, "The Work of Wearing Cameras: Body-Worn Devices and Police Media Labor," in *The Routledge Companion to Labor and Media*, ed. R. Maxwell (New York: Routledge, 2016), 252–64.

25. Brandon Soderberg, "Persistent Transparency: Baltimore Surveillance Plane Documents Reveal Ignored Pleas to Go Public," *Baltimore City Paper*, November 1, 2016; Soderberg, "Baltimore Defense Attorneys."

26. Emphasis added.

27. Benjamin Snyder, "'All We See Is Dots': Aerial Objectivity and Mass Surveillance in Baltimore," *History of Photography* 45, no. 3–4 (2021): 376–87.

28. Louis-Georges Schwartz, *Mechanical Witness: A History of Motion Picture Evidence in U.S. Courts* (New York: Oxford University Press, 2009).

29. Bryce Clayton Newell, *Police Visibility: Privacy, Surveillance, and the False Promise of Body-Worn Cameras* (Berkeley: University of California Press, 2021).

30. Mark Reutter and Fern Shen, "'House Cats' and 'G-Days': A Look at the Overtime Culture at the Baltimore Police Department," *Baltimore Brew*, February

20, 2018, https://www.baltimorebrew.com/2018/02/20/house-cats-and-g-days-a-look-at-overtime-culture-at-the-baltimore-police-department/.

31. My own work on this issue, which was conducted prior to the program's launch, anticipated many of the difficulties PSS would have in achieving an effective sousveillance function. See Snyder, "'Big Brother's Bigger Brother'," 1315–36.

32. Newell, *Police Visibility*.

33. WBEZ Chicago, *Motive* (podcast), season 5, January–March 2023, https://www.wbez.org/shows/motive/.

34. J. Brian Charles, "The Human Toll of Keeping Baltimore Safe," *The Trace*, March 3, 2022, https://www.thetrace.org/2022/03/baltimore-safe-streets-shootings-gun-violence-mayor-scott/.

35. Andrew R. Morral et al., *Evaluating Baltimore/s Aerial Investigation Research Pilot Program: Interim Report* (Santa Monica, CA: RAND, 2021), https://www.rand.org/pubs/research_reports/RRA1131-2.html.

36. J. Cavanaugh Simpson, "Prying Eyes: Military-Grade Surveillance Keeps Watch Over Baltimore and City Protests, but Catches Few Criminals," *Baltimore Magazine*, August 5, 2020, https://www.baltimoremagazine.com/section/community/surveillance-planes-watch-over-baltimore-but-catch-few-criminals/.

Chapter 6

1. Michel Foucault, *Discipline and Punish: The Birth of the Prison*, trans. Alan Sheridan (New York: Pantheon Books, 1977).

2. Leaders of a Beautiful Struggle, Erricka Bridgeford, Kevin James v. Baltimore Police Department, Michael S. Harrison, No. 20-1495 (4th Cir. Ct. App. July 24, 2021), 32.

3. California v. Ciraolo, No. 84-1513 U.S. (May 19, 1986); Dow Chemical Co. v. United States, No. 84-1259 U.S. (May 19, 1986).

4. Baltimore Police Department, *Professional Services Agreement: Aerial Investigation Research ("AIR")* (Baltimore: Baltimore Police Department, March 17, 2020), 19, https://www.baltimorepolice.org/sites/default/files/General%20Website%20PDFs/MOU_AIR_Presented_to_Board_of_Estimates-compressed.pdf. (Emphasis added.)

5. Baltimore Police Department, 23. (Emphasis added.)

6. Neil Richards, *Why Privacy Matters* (New York: Oxford University Press, 2021), 82.

7. The actual language from the MOU: "In addition to... Target Crimes, BPD may request... support in extraordinary and exigent circumstances, on a case by case basis, only as identified and specifically approved in writing by the Baltimore Police Commissioner. Exigent cases may include for example, cases involving imminent danger or loss of life, a kidnapping, chemical spill or train derailment." Baltimore Police Department, *Professional Services Agreement*, 18.

8. Ari Ezra Waldman, *Industry Unbound: The Inside Story of Privacy, Data, and Corporate Power* (Cambridge: Cambridge University Press, 2021), 10. See also Lauren B. Edelman, "Legal Ambiguity and Symbolic Structures: Organizational Mediation of Civil Rights Law," *American Journal of Sociology* 97, no. 6 (1992): 1532.

9. Andrew Guthrie Ferguson, "Persistent Surveillance," *Alabama Law Review*, 2022, https://ssrn.com/abstract = 4071189.

10. Jeffrey Pfeffer and Gerald R. Salancik, *The External Control of Organizations: A Resource Dependence Perspective* (Stanford, CA: Stanford University Press, 2003).

11. Ferguson, "Persistent Surveillance," 52.

12. David R. Rocah et al., Leaders of a Beautiful Struggle, Erricka Bridgeford, Kevin James v. Baltimore Police Department, Michael S. Harrison—Complaint for Declaratory and Injunctive Relief, No. 20-929, (Dis. Ct. of Md. April 9, 2020), 2.

13. *Rocah et al.* at 6.

14. In the first Fourth Circuit hearing, the court upheld the lower court's ruling that the plane did not violate the Fourth Amendment. They then overturned their own ruling in the second, *en banc*, hearing.

15. The dissenting opinion, authored by Wilkinson, had little to say about the issue of privacy. The main concern was that the circuit court had overturned a lower court, which, he argued, was better at representing what the people of Baltimore actually wanted. He cites, for example, PSS's extensive community outreach campaign and public opinion polling as illustrating public support for the spy plane. A higher court, he argued, should not bureaucratically tell Baltimoreans what they should want. See Cameron Chiani, "Leaders of a Beautiful Struggle v. Baltimore Police Department: Balancing the Advances of Police Tracking with the Constitutional Rights Afforded to the Public Citizenry," *Journal of Business and Technology Law* 17, no. 2 (2022): 385.

16. *Leaders of a Beautiful Struggle*, No. 20-1495 at 21.

17. Christopher Slobogin, *Virtual Searches: Regulating the Covert World of Technological Policing* (Oakland: University of California Press, 2022).

18. Carpenter v. United States, No. 16-402 U.S. (June 22, 2018) 2.

19. The dissenting opinions disagreed on this point. Gorsuch, for example, noted that the phrase "reasonable expectation" has never been clearly defined and, however we define it, it should hardly be left up to federal judges to define on behalf of citizens. Reasonableness should be left up to public opinion and/or local legislators in conversation with their constituents. See *Carpenter*, No. 16-402 at 8, Gorsuch, J., dissenting.

20. Orin Kerr has called this the "mosaic theory" of the Fourth Amendment: an unconstitutional search can arise from an aggregate of smaller constitutional searches. Orin S. Kerr, "The Mosaic Theory of the Fourth Amendment," *Michigan Law Review* 111, no. 3 (2012): 311–54.

21. Andrew Guthrie Ferguson, "Facial Recognition and the Fourth Amendment," *Minnesota Law Review* 105 (2021): 1138.

22. *Carpenter*, No. 16-402 at 11, Gorsuch, J., dissenting.

23. *Carpenter* at 2.

24. Paul Ohm, "The Many Revolutions of Carpenter," *Harvard Journal of Law and Technology* 32, no. 2 (2019): 399.

25. Ohm, 403.

26. Ohm, 402.

27. *Leaders of a Beautiful Struggle*, No. 20-1495 at 19.

28. *Leaders of a Beautiful Struggle* at 32.

29. *Rocah et al.*, No. 20-929 at 2.

30. Lee Vinsel, "You're Doing It Wrong: Notes on Criticism and Technology Hype," *STS News* (blog), February 1, 2021, https://sts-news.medium.com/youre-doing-it-wrong-notes-on-criticism-and-technology-hype-18b08b4307e5.

31. Slobogin, *Virtual Searches*, 201.

32. Julie E. Cohen, *Configuring the Networked Self: Law, Code, and the Play of Everyday Practice* (New Haven, CT: Yale University Press, 2012), 239.

33. "AI Hype Distracted Us from Real Problems, with Timnit Gebru," *Tech Won't Save Us* (podcast), January 18, 2024, https://techwontsave.us/episode/203_ai_hype_distracted_us_from_real_problems_w_timnit_gebru.

Chapter 7

1. Mosby did not reply to requests for a comment.

2. The prosecutors in the Carroll Avenue case did not respond to requests to be interviewed.

3. Rogers lamented the way this decision was made. "This wasn't even done in an open hearing," she said. It was just an informal decision made by the judge in an off-the-record conversation. I was never able to find public footage, audio, or any documentation of why the judge threw out the spy plane evidence in the Carroll Avenue case, or any other. I attempted to speak with the judge about her unofficial ruling, but received conflicting information. Through a public relations official, she informed me that both the McHenry and Carroll cases had been voluntarily dismissed by prosecutors (*nolle prosequi*), and so she "did not rule on those cases." I don't know who is telling the truth here. Was the judge unconvinced by the argument put forward by prosecutors that police acted in good faith? Or, did prosecutors dismiss the cases before the judge needed to rule?

4. The private attorney representing this defendant never filed a motion on behalf of his client to dismiss the case or suppress the spy plane evidence, even after the Fourth Circuit decision had come down. "In my experience," he told me, "the prosecution's case doesn't get any better with age." Feeling no urgency to push for a speedy result for his client, this attorney allowed the defendant to sit in jail on the faith that prosecutors would eventually give up on their case because it was just too hard to prosecute.

5. I did not speak directly to the suspects, victims, or victims' families who were involved in the prosecuted spy plane cases. Talking to suspects could have risked self-incrimination and was, therefore, never a consideration. Talking to victims and victims' families involved delicate ethical considerations, which I explore at length in the appendix. In terms of understanding how the dismissed spy plane cases affected victims and their families, then, all I have to go on is what police and attorneys told me. While these stories are secondhand, and therefore their accuracy cannot be verified, I think this is still the *least* unethical way of relating this important information. I also acknowledge and regret that there is still something extractive about conveying this information without full communication with the families.

6. Weihua Li and Jamiles Lartey, "As Murders Spiked, Police Solved about Half in 2020," *Marshall Project* (blog), January 12, 2022, https://www.themarshallproject.org/2022/01/12/as-murders-spiked-police-solved-about-half-in-2020.

7. Christopher T. Lowenkamp, Marie VanNostrand, and Alexander Holsinger, *The Hidden Costs of Pretrial Detention* (Houston: Laura and John Arnold Foundation, 2013).

8. Louis-Georges Schwartz, *Mechanical Witness: A History of Motion Picture Evidence in U.S. Courts* (New York: Oxford University Press, 2009).

9. Quoted in Charles Goodwin, "Professional Vision," *American Anthropologist* 96, no. 3 (1994): 615.

10. Goodwin, 606.

11. Goodwin, 616.

12. Susan Schuppli, *Material Witness: Media, Forensics, Evidence* (Cambridge, MA: MIT Press, 2020).

13. Mary D. Fan, "Body Cameras, Big Data, and Police Accountability," *Law and Social Inquiry* 43, no. 4 (2018): 1236–56; Bryce Clayton Newell, "Context, Visibility, and Control: Police Work and the Contested Objectivity of Bystander Video," *New Media and Society* 21, no. 1 (2019): 60–76; Bryce Clayton Newell, *Police Visibility: Privacy, Surveillance, and the False Promise of Body-Worn Cameras* (Berkeley: University of California Press, 2021).

14. Sarah Brayne, Karen Levy, and Bryce Clayton Newell, "Visual Data and the Law," *Law and Social Inquiry* 43, no. 4 (2018): 1150.

15. Forrest Stuart, "Constructing Police Abuse after Rodney King: How Skid Row Residents and the Los Angeles Police Department Contest Video Evidence," *Law and Social Inquiry* 36, no. 2 (2011): 327–53; Kelly Gates, "The Cultural Labor of Surveillance: Video Forensics, Computational Objectivity, and the Production of Visual Evidence," *Social Semiotics* 23, no. 2 (2013): 242–60; Kelly Gates, "The Work of Wearing Cameras: Body-Worn Devices and Police Media Labor," in *The Routledge Companion to Labor and Media*, ed. R. Maxwell (New York: Routledge, 2016), 252–64; Emmeline Taylor, "Lights, Camera, Redaction . . . Police Body-Worn Cameras: Autonomy, Discretion, and Accountability," *Surveillance and Society* 14, no. 1 (2016): 128–32.

16. Quoted in Brayne, Levy, and Newell, "Visual Data and the Law," 1150.

17. Shaundria's criticism didn't seem to convince McNutt and other program leaders. McNutt pushed back that "tracking everything is part of our process. How can you know who's involved and who's not until you've tracked it all?" He continued to instruct analysts to track as many targets as they could. Shaundria, feeling disregarded and underutilized for her expertise, ultimately left the company before the 2020 trial ended.

18. John Shiffman and Kristina Cooke, "U.S. Directs Agents to Cover Up Program Used to Investigate Americans," *Reuters*, August 5, 2013, https://www.reuters.com/article/us-dea-sod-idUSBRE97409R20130805.

19. Human Rights Watch, *Dark Side: Secret Origins of Evidence in US Criminal Cases* (New York: Human Rights Watch, 2018).

20. Karen McVeigh, "US Drug Agency Surveillance Unit to Be Investigated by Department of Justice," *The Guardian*, August 6, 2013, sec. World

21. Human Rights Watch, *Dark Side*, 39.

22. Human Rights Watch, 50.

23. Brian Pori and Sarah St. Vincent, *Parallel Construction: How to Discover the Government's Undisclosed Sources of Evidence* (webinar), National Association of Criminal Defense Lawyers, May 23, 2018, https://www.nacdl.org/Media/Parallel-Construction-Discover-Govt-Evidenc-Source.

24. Janet Vertesi, *Seeing Like a Rover: How Robots, Teams, and Images Craft Knowledge of Mars* (Chicago: University of Chicago Press, 2014); Beth A. Bechky, *Blood, Powder, and Residue: How Crime Labs Translate Evidence into Proof* (Princeton, NJ: Princeton University Press, 2021).

25. Zeynep Devrim Gürsel, *Image Brokers: Visualizing World News in the Digital Age of Circulation* (Berkeley: University of California Press, 2016).

Chapter 8

1. Theo Wayt, "Citizen Pays New Yorkers $25 an Hour to Livestream Crime Scenes," *New York Post*, July 25, 2021, https://nypost.com/2021/07/25/citizen-pays-new-yorkers-25-an-hour-to-livestream-crime-scenes/.

2. Lee Vinsel, "You're Doing It Wrong: Notes on Criticism and Technology Hype," *STS News* (blog), February 1, 2021, https://sts-news.medium.com/youre-doing-it-wrong-notes-on-criticism-and-technology-hype-18b08b4307e5.

3. Jenna Burrell, "Artificial Intelligence and the Ever-Receding Horizon of the Future," *Tech Policy Press*, June 6, 2023, https://techpolicy.press/artificial-intelligence-and-the-ever-receding-horizon-of-the-future/.

4. The criminal legal system is one of the primary sites of experimentation for AI. Jameson Spivack, *Cop Out: Automation in the Criminal Legal System* (Washington, DC: Georgetown Law Center on Privacy and Technology, March 23, 2023), https://copout.tech/wp-content/uploads/2023/03/CPT_Cop-Out_Essay.pdf.

5. "Statement on AI Risk: AI Experts and Public Figures Express Their Concern about AI Risk," Center for AI Safety, March 30, 2023, https://www.safe.ai/statement-on-ai-risk.

6. Matteo Wong, "AI Doomerism Is a Decoy," *The Atlantic*, June 2, 2023, https://www.theatlantic.com/technology/archive/2023/06/ai-regulation-sam-altman-bill-gates/674278/.

7. Catherine O'Neil, *Weapons of Math Destruction: How Big Data Increases Inequality and Threatens Democracy* (New York: Crown Random House, 2016); Virginia Eubanks, *Automating Inequality: How High-Tech Tools Profile, Police, and Punish the Poor* (New York: St. Martin's Press, 2017); Joy Boulamwini and Timnit

Gebru, "Gender Shades: Intersectional Accuracy Disparities in Commercial Gender Classification," *Proceedings of Machine Learning Research* 81 (2018): 1–15; Ruha Benjamin, *Race after Technology: Abolitionist Tools for the New Jim Code* (New York: Polity, 2019); Meredith Broussard, *Artificial Unintelligence: How Computers Misunderstand the World* (Cambridge, MA: MIT Press, 2019); Sarah Brayne, *Predict and Surveil: Data, Discretion, and the Future of Policing* (New York: Oxford University Press, 2020); Emily Bender et al., "On the Dangers of Stochastic Parrots: Can Language Models Be Too Big?," in *FAccT '21: Proceedings of the 2021 ACM Conference on Fairness, Accountability, and Transparency*, March 2021, 610–23, https://doi.org/10.1145/3442188.3445922; Joy Boulamwini, *Unmasking AI: My Mission to Protect What Is Human in a World of Machines* (New York: Random House, 2023); Meredith Broussard, *More than a Glitch: Confronting Race, Gender, and Ability Bias in Tech* (Cambridge, MA: MIT Press, 2023).

8. R. Joshua Scannell, "This Is Not Minority Report: Predictive Policing and Population Racism," in *Captivating Technology: Race, Carceral Technosciene, and Liberatory Imagination in Everyday Life*, ed. Ruha Benjamin (Durham, NC: Duke University Press, 2019), 107–29.

9. Brayne, *Predict and Surveil*.

10. See also Aaron Sankin et al., "Crime Prediction Software Promised to Be Free of Biases; New Data Shows It Perpetuates Them," *The Markup*, December 2, 2021.

11. Leila Miller, "LAPD Will End Controversial Program That Aimed to Predict Where Crimes Would Occur," *Los Angeles Times*, April 21, 2020, https://www.latimes.com/california/story/2020-04-21/lapd-ends-predictive-policing-program.

12. Mark Puente, "LAPD Pioneered Predicting Crime with Data; Many Police Don't Think It Works," *Los Angeles Times*, July 3, 2019, https://www.latimes.com/local/lanow/la-me-lapd-precision-policing-data-20190703-story.html.

13. Sarah Brayne, "Big Data Surveillance: The Case of Policing," *American Sociological Review* 82, no. 8 (2017): 977–1008.

14. Brayne relays a story from a detective in the LAPD's Juvenile Division who gushed about using Palantir, in combination with automatic license plate readers, to set up a geofence around a school in order to catch an alleged child predator. The technology helped them track six cars that had all frequented the area around the school, but none of them ended up matching the car of their original suspect. The detective ultimately admitted that the geofencing operation did not result in an arrest. Brayne, *Predict and Surveil*, 51–52.

15. ShotSpotter denies these findings, claiming that their own numbers indicate a nearly 99 percent accuracy rate. Their number, however, is based on self-reporting by police officers, who, you can imagine, probably do not bother to call up the company when an alert turns up nothing. The 99 percent rate most likely excludes the vast majority of the system's false positives. Lucy Parsons Labs, "ShotSpotter Creates Thousands of Unfounded Police Deployments, Fuels Unconstitutional Stop-and-Frisk, and Can Lead to False Arrests," MacArthur Justice Center, July 21, 2022, https://endpolicesurveillance.com/.

16. Joseph M. Ferguson and Deborah Witzburg, *The Chicago Police Department's Use of ShotSpotter Technology* (Chicago: Office of Inspector General, August 2021).

17. Jamie Kalven, "Chicago Awaits Video of Police Killing of 13-Year-Old Boy," *The Intercept*, April 13, 2021, https://theintercept.com/2021/04/13/chicago-police-killing-boy-adam-toledo-shotspotter/.

18. Garance Burke and Michael Tarm, "Confidential Document Reveals Key Human Role in Gunshot Tech," AP, January 20, 2023, https://apnews.com/article/shotspotter-artificial-intelligence-investigation-9cb47bbfb565dc3ef-110f92ac7f83862.

19. Office of the Mayor, "City of Chicago Statement on ShotSpotter Contract," Chicago, Mayor's Press Office, February 13, 2024, https://www.chicago.gov/city/en/depts/mayor/press_room/press_releases/2024/january/city-of-chicago-statement-on-shotspotter-contract.html.

20. Kashmir Hill, *Your Face Belongs to Us: A Secretive Startup's Quest to End Privacy as We Know It* (New York: Penguin Random House, 2023).

21. Boulamwini and Gebru, "Gender Shades," 1–15.

22. Miriam Marini, "Farmington Hills Man Sues Detroit Police after Facial Recognition Wrongly Identifies Him," *Detroit Free Press*, April 13, 2021, https://www.freep.com/story/news/local/michigan/2021/04/13/detroit-police-wrongful-arrest-faulty-facial-recognition/7207135002/.

23. Boulamwini, *Unmasking AI*.

24. Clare Garvie, "Garbage In, Garbage Out: Face Recognition on Flawed Data," Georgetown Law Center on Privacy and Technology, May 16, 2019, https://www.flawedfacedata.com/.

25. This kind of corner cutting is commonly observed in studies of computer-human interaction in other work settings. See Min Kyung Lee et al., "Working with Machines: The Impact of Algorithmic and Data-Driven Management on Human Workers," in *Proceedings of the 33rd Annual ACM Conference on Human*

Factors in Computing Systems, ed. B. Begole et al. (New York: ACM Press, 2015), 1603–12.

26. Robert K. Merton, "The Unanticipated Consequences of Purposive Social Action," *American Sociological Review* 1, no. 6 (1936): 894–904.

27. Jeffrey Pfeffer and Gerald R. Salancik, *The External Control of Organizations: A Resource Dependence Perspective* (Stanford, CA: Stanford University Press, 2003).

28. Alvaro M. Bedoya, "The Color of Surveillance: What an Infamous Abuse of Power Teaches Us about the Modern Spy Era," *Slate*, January 18, 2016, https://slate.com/technology/2016/01/what-the-fbis-surveillance-of-martin-luther-king-says-about-modern-spying.html.

29. Brayne, *Predict and Surveil*, 141.

30. Brayne, 142.

31. Zana Buçinca et al., *AHA!: Facilitating AI Impact Assessment by Generating Examples of Harms* (Ithaca, NY: Cornell University, arXiv, June 2023), https://doi.org/arXiv:2306.03280v1.

32. "Community Control Over Police Surveillance," American Civil Liberties Union, 2020, https://www.aclu.org/issues/privacy-technology/surveillance-technologies/community-control-over-police-surveillance.

33. Linda L. Dahlberg and James A. Mercy, "The History of Violence as a Public Health Issue," *AMA Virtual Mentor* 11, no. 2 (2009): 167–72, https://journalofethics.ama-assn.org/article/history-violence-public-health-problem/2009-02; Anthony A. Braga and David L. Weisburd, "The Effects of Focused Deterrence Strategies on Crime: A Systematic Review and Meta-Analysis of the Empirical Evidence," *Journal of Research in Crime and Delinquency* 49, no. 3 (2012): 323–58; Jeffrey A. Butts et al., "Cure Violence: A Public Health Model to Reduce Gun Violence," *Annual Review of Public Health* 36 (2015): 39–53.

34. Butts et al., "Cure Violence," 39–53.

35. Office of the Mayor, *Gun Violence Reduction Task Force—Introductory Memo* (New Orleans: City of New Orleans, August 13, 2018), https://www.documentcloud.org/documents/5017837-Intro-Memo-Gun-Violence-Reduction-Task-Force.html#document/p4/a462954.

36. Thomas Abt, *Bleeding Out: The Devastating Consequences of Urban Violence—and a Bold New Plan for Peace in the Streets* (New York: Basic Books, 2019); Thomas Ward Frampton, "The Dangerous Few: Taking Seriously Prison Abolition and Its Skeptics," *Harvard Law Review* 135, no. 8 (2022): 2013–52.

37. WBEZ Chicago, *Motive* (podcast), season 5, January–March 2023, https://www.wbez.org/shows/motive/.

38. Mark Andrejevic, "The Work of Watching One Another: Lateral Surveillance, Risk, and Governance," *Surveillance and Society* 2, no. 4 (2004): 479–97.

39. The option to decline the intervention is possibly the most controversial aspect of public health interventions. The focused deterrence model, for example, takes a more aggressive tack here, suggesting that high-risk individuals should face punishment if they decline to take the assistance offered to them. See Braga and Weisburd, "Effects of Focused Deterrence Strategies on Crime," 323–58.

40. Heartland Alliance, *Working Together toward Safer Communities: Reflections from READI Chicago* (Chicago: Heartland Alliance, 2022), https://www.heartlandalliance.org/wp-content/uploads/2021/09/READI-Chicago-Working-Together-Toward-Safer-Communities-small.pdf.

41. Christopher Blattman et al., "Cognitive Behavior Therapy Reduces Crime and Violence over 10 Years: Experimental Evidence," *SocArXiv*, May 16, 2022, https://doi.org/10.31235/osf.io/q85ux.

42. J. Brian Charles, "The Human Toll of Keeping Baltimore Safe," *The Trace*, March 3, 2022, https://www.thetrace.org/2022/03/baltimore-safe-streets-shootings-gun-violence-mayor-scott/.

43. WBEZ Chicago, *Motive*.

44. Nick Chrastil and Katy Reckdahl, "A Group of 'Violence Interrupters' Worked the Streets of New Orleans to Prevent Retaliatory Shootings—until They Were Sidelined 2 Years Ago," *The Lens* (blog), May 25, 2023, https://thelensnola.org/2023/05/25/a-group-of-violence-interrupters-worked-the-streets-of-new-orleans-to-prevent-retaliatory-shootings-until-they-were-sidelined-2-years-ago/.

45. Sarah Brayne, "Surveillance and System Avoidance: Criminal Justice Contact and Institutional Attachment," *American Sociological Review* 79, no. 3 (2014): 367–91.

46. James H. Jones, *Bad Blood: The Tuskegee Syphilis Experiment*, rev. ed. (New York: Free Press, 1993); Harriet A. Washington, *Medical Apartheid: The Dark History of Medical Experimentation on Black Americans from Colonial Times to the Present* (New York: Vintage, 2008).

47. In 2022, McNutt tapped Joyous Jones and some of the other backers of the original program to help lobby the state government to increase the reward amounts. They succeeded, increasing rewards for tips on gun-related crime to $8,000.

48. The RAND study showed that the AIR program, though small in its effect size, had greater success at helping solve carjacking cases than other crime types. Andrew R. Morral et al., *Preliminary Findings from the Aerial Investigations Research Pilot Program* (Santa Monica, CA: RAND, January 27, 2021), 3.

Appendix

1. W. E. B. Du Bois, "Of Our Spiritual Strivings," in *The Souls of Black Folk* (Chicago: A. C. McLurg, 1903), chap. 1.

2. Laura Nader, "Up the Anthropologist: Perspectives Gained from Studying Up," ERIC Clearinghouse, 1972, https://eric.ed.gov/?id = ED065375; Nick Seaver, "Studying Up: The Ethnography of Technologists," *Ethnography Matters* (blog), March 10, 2014, https://ethnographymatters.net/blog/2014/03/10/studying-up/.

3. For outstanding examples of ethnographic work documenting the experiences of being watched, see Victor M. Rios, *Punished: Policing the Lives of Black and Latino Boys* (New York: NYU Press, 2011); Randol Contreras, *The Stickup Kids: Race, Drugs, Violence, and the American Dream* (Oakland: University of California Press, 2013); Forrest Stuart, *Down, Out, and Under Arrest: Policing and Everyday Life in Skid Row* (Chicago: University of Chicago Press, 2016); Asad L. Asad, *Engage and Evade: How Latino Immigrant Families Manage Surveillance in Everyday Life* (Princeton, NJ: Princeton University Press, 2023); Melissa Villa-Nicholas, *Data Borders: How Silicon Valley Is Building an Industry around Immigrants* (Oakland: University of California Press, 2023).

4. Steven Lubet, *Interrogating Ethnography: Why Evidence Matters* (New York: Oxford University Press, 2018).

5. I am deeply grateful to Christina Simko and sociology students at Williams College for helping me think through this decision in advance of the fieldwork.

6. Barry Friedman, Farhang Heydari, and Max Isaacs, *Brief of the Policing Project as Amicus Curiae in Support of Neither Party and in Support of Rehearing or Rehearing En Banc* (New York: NYU Policing Project, November 27, 2020), 6.

7. Barry Friedman et al., *Civil Rights and Civil Liberties Audit of Baltimore's Aerial Investigation Research (AIR) Program* (New York: Policing Project at NYU Law, 2020), 21, https://www.policingproject.org/s/AIR-Program-Audit-Report-vFINAL-reduced.pdf.

8. John Van Maanen, "The Moral Fix: On the Ethics of Fieldwork," in *Contemporary Field Research: A Collection of Readings* (Boston: Little Brown, 1983), 269–87; Rik Scarce, "(No) Trial (but) Tribulations: When Courts and Ethnography Conflict," *Journal of Contemporary Ethnography* 23, no. 2 (1994): 123–49; Richard A. Leo, "Trials and Tribulations: Courts, Ethnography, and the Need for an Evidentiary Privilege for Academic Researchers," *American Sociologist* 26 (1995): 113–34; Shamus Khan, "The Subpoena of Ethnographic Data," *Sociological Forum* 34, no. 1 (2019): 253–63; Jack Katz, "Armor for Ethnographers," *Sociological Forum* 34, no. 1 (2019): 264–75.

9. Ted Palys and John Lowman, "Defending Research Confidentiality 'to the Extent the Law Allows': Lessons from the Boston College Subpoenas," *Journal of Academic Ethics* 10 (2012): 271–97.

10. Khan, "Subpoena of Ethnographic Data," 253–63.

11. Scarce, "(No) Trial (but) Tribulations," 123–49.

12. Ted Palys and John Lowman, *The Belfast Project Autopsy: Who Can You Trust?* (London: Routledge, 2016); Andres Ruiz, "The Belfast Project: How an American University Almost Started a Civil War," *StMU Research Scholars* (blog), November 30, 2021, https://stmuscholars.org/the-belfast-project-how-an-american-university-almost-started-a-civil-war/.

13. Palys and Lowman, "Defending Research Confidentiality," 271–97.

14. Michael A. Cusumano and David B. Yoffie, *Competing on Internet Time: Lessons from Netscape and Its Battle with Microsoft* (New York: Free Press, 2000).

15. Cusumano v. Microsoft Corporation, No. 98–2133 (U.S. 1st Cir. Ct. App. Dec. 15, 1998), 29.

16. This level of protection has been offered even in a criminal investigation. In the 1980s, a University of Albany sociology PhD student was privy to a potential arson in a restaurant where he was conducting ethnography. He had his field notes subpoenaed. He successfully quashed the request because, again, the judge compared his privilege of academic freedom to the "qualified privilege" of journalists to protect the free flow of information to the public. See Jack B. Weinstein and Catherine Wimberly, "Secrecy in Law and Science," *Cardozo Law Review* 23, no. 1 (2001): 12.

17. Ted Palys and John Lowman, *Informed Consent, Confidentiality and the Law: Implications of the Tri-Council Policy Statement* (Burnaby, BC: Simon Fraser University, January 31, 1999), http://www.sfu.ca/~palys/Conf&Law.html Appendix A.

18. Ted Palys and John Lowman, "Anticipating Law: Research Methods, Ethics, and the Law of Privilege," *Sociological Methodology* 32 (2002): 1–17. For a critical view of this advice, see Geoffrey R. Stone, "Discussion—Above the Law: Research Methods, Ethics, and the Law of Privilege," *Sociological Methodology* 32 (2002): 19–27.

19. The Wigmore test has four criteria that must be met in order to balance confidentiality above disclosure. For a lengthy discussion, see Palys and Lowman, *Informed Consent, Confidentiality and the Law*.

20. Ted Palys and John Lowman, *Protecting Research Confidentiality: What Happens When Law and Ethics Collide* (Toronto: James Lorimer, 2014).

21. Katz, "Armor for Ethnographers," 274.

Selected Bibliography

The list below contains key books, articles, and reports that are also referenced in the notes for each chapter. We academics love long bibliographies, but for a book that is meant for a popular audience I thought it would be better to simplify this section as much as possible, hopefully easing readers' ability to find the sources that will truly support them in going deeper. I have omitted from this list most short newspaper articles, blog posts, and podcast episodes that, in my opinion, didn't rise to the level of essential reading on the topic of the spy plane and crime technology. These sources can still be found in the full notes for each chapter.

Abt, Thomas. *Bleeding Out: The Devastating Consequences of Urban Violence—and a Bold New Plan for Peace in the Streets.* New York: Basic Books, 2019.

"AI Hype Distracted Us from Real Problems, with Timnit Gebru." *Tech Won't Save Us* (podcast), January 18, 2024. https://techwontsave.us/episode/203_ai_hype_distracted_us_from_real_problems_w_timnit_gebru.

Alpert, Geoffrey P., Jeffrey J. Noble, and Jeff Rojek. "Solidarity and the Code of Silence." In *Critical Issues in Policing: Contemporary Readings*, edited by Roger G. Dunham, Geoffrey P. Alpert, and Kyle D. McLean, 106-21. Long Grove, IL: Waveland Press, 2015.

Anderson, Elijah. *Code of the Street: Decency, Violence, and the Moral Life of the Inner City.* New York: W. W. Norton, 1999.

Andrejevic, Mark. "The Work of Watching One Another: Lateral Surveillance, Risk, and Governance." *Surveillance and Society* 2, no. 4 (2004): 479-97.

Arsiniega, Brittany, and Matthew Guariglia. "Police as Supercitizens." *Social Justice* 48, no. 4 (2021): 33-58.

Asad, Asad L. *Engage and Evade: How Latino Immigrant Families Manage Surveillance in Everyday Life.* Princeton, NJ: Princeton University Press, 2023.

Bechky, Beth A. *Blood, Powder, and Residue: How Crime Labs Translate Evidence into Proof.* Princeton, NJ: Princeton University Press, 2021.

Bell, Monica C. "Anti-Segregation Policing." *New York University Law Review* 95 (2020): 650–765.

Bender, Emily, Timnit Gebru, Angelina McMillan-Major, and Shmargaret Shmitchell. "On the Dangers of Stochastic Parrots: Can Language Models Be Too Big?" In *FAccT '21: Proceedings of the 2021 ACM Conference on Fairness, Accountability, and Transparency*, March 2021, 610–23. https://doi.org/10.1145/3442188.3445922.

Benjamin, Ruha. *Captivating Technology: Race, Carceral Technoscience, and Liberatory Imagination in Everyday Life.* Durham, NC: Duke University Press, 2019.

———. *Race after Technology: Abolitionist Tools for the New Jim Code.* New York: Polity, 2019.

Blattman, Christopher, Sebastian Chaskel, Julian C. Jamison, and Margaret Sheridan. "Cognitive Behavior Therapy Reduces Crime and Violence over 10 Years: Experimental Evidence." *SocArXiv*, May 16, 2022. https://doi.org/10.31235/osf.io/q85ux.

Bonilla-Silva, Eduardo. *Racism without Racists: Color-Blind Racism and the Persistence of Racial Inequality in America.* 6th ed. New York: Rowman and Littlefield, 2021.

Boulamwini, Joy. *Unmasking AI: My Mission to Protect What Is Human in a World of Machines.* New York: Random House, 2023.

Boulamwini, Joy, and Timnit Gebru. "Gender Shades: Intersectional Accuracy Disparities in Commercial Gender Classification." *Proceedings of Machine Learning Research* 81 (2018): 1–15.

Braga, Anthony A., and David L. Weisburd. "The Effects of Focused Deterrence Strategies on Crime: A Systematic Review and Meta-Analysis of the Empirical Evidence." *Journal of Research in Crime and Delinquency* 49, no. 3 (2012): 323–58.

Brayne, Sarah. "Big Data Surveillance: The Case of Policing." *American Sociological Review* 82, no. 8 (2017): 977–1008.

———. *Predict and Surveil: Data, Discretion, and the Future of Policing.* New York: Oxford University Press, 2020.

———. "Surveillance and System Avoidance: Criminal Justice Contact and Institutional Attachment." *American Sociological Review* 79, no. 3 (2014): 367–91.

Brayne, Sarah, and Angele Christin. "Technologies of Crime Prediction: The Reception of Algorithms in Policing and Criminal Courts." *Social Problems*, 2020. https://doi.org/10.1093/socpro/spaa004.

Brayne, Sarah, Karen Levy, and Bryce Clayton Newell. "Visual Data and the Law." *Law and Social Inquiry* 43, no. 4 (2018): 1149–63.

Bromwich, Michael R., Jason M. Weinstein, Rachel B. Peck, Katherine M. Dubyak, William G. Fletcher, and James M. Purce. *Anatomy of the Gun Trace Task Force Scandal: Its Origins, Causes, and Consequences*. Washington, DC: Steptoe, 2022.

Broussard, Meredith. *Artificial Unintelligence: How Computers Misunderstand the World*. Cambridge, MA: MIT Press, 2019.

———. *More than a Glitch: Confronting Race, Gender, and Ability Bias in Tech*. Cambridge, MA: MIT Press, 2023.

Brown, Lawrence T. *The Black Butterfly: The Harmful Politics of Race and Space in America*. Baltimore: Johns Hopkins University Press, 2021.

Browning, John. "#Snitches Get Stitches: Witness Intimidation in the Age of Facebook and Twitter." *Pace Law Review* 35, no. 1 (2014): 192–214.

Buçinca, Zana, Chau Minh Pham, Maurice Jakesch, Marco Tulio Ribeiro, Alexandra Olteanu, and Saleema Amershi. *AHA!: Facilitating AI Impact Assessment by Generating Examples of Harms*. Ithaca, NY: Cornell University, arXiv, June 2023. https://doi.org/arXiv:2306.03280v1.

Burrell, Jenna. "Artificial Intelligence and the Ever-Receding Horizon of the Future." *Tech Policy Press*, June 6, 2023. https://techpolicy.press/artificial-intelligence-and-the-ever-receding-horizon-of-the-future/.

Butts, Jeffrey A., Caterina Gouvis Roman, Lindsay Bostwick, and Jeremy R. Porter. "Cure Violence: A Public Health Model to Reduce Gun Violence." *Annual Review of Public Health* 36 (2015): 39–53.

California v. Ciraolo, No. 84-1513 U.S. (May 19, 1986).

Carpenter v. United States, No. 16-402 U.S. (June 22, 2018).

Center for Investigative Reporting. "Is ShotSpotter Worth the Expense?" *Reveal* (podcast). Accessed June 12, 2023. https://on.soundcloud.com/wXsVs.

Cerulo, Karen A. *Never Saw It Coming: Cultural Challenges to Envisioning the Worst*. Chicago: University of Chicago Press, 2006.

Charles, J. Brian. "The Human Toll of Keeping Baltimore Safe." *The Trace*, March 3, 2022. https://www.thetrace.org/2022/03/baltimore-safe-streets-shootings-gun-violence-mayor-scott/.

Chiani, Cameron. "Leaders of a Beautiful Struggle v. Baltimore Police Department: Balancing the Advances of Police Tracking with the Constitutional

Rights Afforded to the Public Citizenry." *Journal of Business and Technology Law* 17, no. 2 (2022): 345–62.

Cohen, Julie E. *Configuring the Networked Self: Law, Code, and the Play of Everyday Practice*. New Haven, CT: Yale University Press, 2012.

Contreras, Randol. *The Stickup Kids: Race, Drugs, Violence, and the American Dream*. Oakland: University of California Press, 2013.

Cotton, Ann P., Gabriela Wasileski, Elias Nader, and Sarah Ficenec. *Baltimore Aerial Investigation Research Project: Findings from the Early Launch Community Survey*. Baltimore: Schaefer Center for Public Policy, University of Baltimore, June 2020. https://68i.ab1.myftpupload.com/wp-content/uploads/2020/12/AIRCommunitySurveyReport-SchaeferCenter-FINAL.pdf.

Cusumano v. Microsoft Corporation, No. 98-2133 (U.S. 1st Cir. Ct. App. Dec. 15, 1998).

Cusumano, Michael A., and David B. Yoffie. *Competing on Internet Time: Lessons from Netscape and Its Battle with Microsoft*. New York: Free Press, 2000.

Dahlberg, Linda L., and James A. Mercy. "The History of Violence as a Public Health Issue." *AMA Virtual Mentor* 11, no. 2 (2009): 167–72. https://journalofethics.ama-assn.org/article/history-violence-public-health-problem/2009-02.

Dow Chemical Co. v. United States, No. 84-1259 U.S. (May 19, 1986).

Du Bois, W. E. B. "Of Our Spiritual Strivings." Chap. 1 in *The Souls of Black Folk*. Chicago: A. C. McLurg, 1903.

Edelman, Lauren B. "Legal Ambiguity and Symbolic Structures: Organizational Mediation of Civil Rights Law." *American Journal of Sociology* 97, no. 6 (1992): 1531–76.

Eubanks, Virginia. *Automating Inequality: How High-Tech Tools Profile, Police, and Punish the Poor*. New York: St. Martin's Press, 2017.

Fan, Mary D. "Body Cameras, Big Data, and Police Accountability." *Law and Social Inquiry* 43, no. 4 (2018): 1236–56.

Fenton, Justin. *We Own This City: A True Story of Crime, Cops, and Corruption*. New York: Random House, 2021.

Ferguson, Andrew Guthrie. "Facial Recognition and the Fourth Amendment." *Minnesota Law Review* 105 (2021): 1105–1209.

———. "Persistent Surveillance." *Alabama Law Review*, 2022. https://ssrn.com/abstract=4071189.

Ferguson, Joseph M., and Deborah Witzburg. *The Chicago Police Department's Use of ShotSpotter Technology*. Chicago: Office of Inspector General, August 2021.

Forman Jr., James. *Locking Up Our Own: Crime and Punishment in Black America.* New York: Farrar, Straus, and Giroux, 2017.

Foucault, Michel. *Discipline and Punish: The Birth of the Prison.* Translated by Alan Sheridan. New York: Pantheon Books, 1977.

Frampton, Thomas Ward. "The Dangerous Few: Taking Seriously Prison Abolition and Its Skeptics." *Harvard Law Review* 135, no. 8 (2022): 2013–52.

Freitag, Douglas, Terry Wohlers, and Therese Philippi. *Rapid Prototyping: State of the Art.* Chicago: Manufacturing Technology Information Analysis Center, October 23, 2003. https://apps.dtic.mil/sti/citations/ADA435248.

Friedersdorf, Conor. "Eyes Over Compton: How Police Spied on a Whole City." *The Atlantic*, April 21, 2014. https://www.theatlantic.com/national/archive/2014/04/sheriffs-deputy-compares-drone-surveillance-of-compton-to-big-brother/360954/.

Friedman, Barry, Farhang Heydari, and Max Isaacs. *Brief of the Policing Project as Amicus Curiae in Support of Neither Party and in Support of Rehearing or Rehearing En Banc.* New York: NYU Policing Project, November 27, 2020.

Friedman, Barry, Farhang Heydari, Emmanuel Mauleón, and Max Isaacs. *Civil Rights and Civil Liberties Audit of Baltimore's Aerial Investigation Research (AIR) Program.* New York: Policing Project at NYU Law, 2020. https://www.policingproject.org/s/AIR-Program-Audit-Report-vFINAL-reduced.pdf.

Garvie, Clare. "Garbage In, Garbage Out: Face Recognition on Flawed Data." Georgetown Law Center on Privacy and Technology, May 16, 2019. https://www.flawedfacedata.com/.

Gates, Kelly. "The Cultural Labor of Surveillance: Video Forensics, Computational Objectivity, and the Production of Visual Evidence." *Social Semiotics* 23, no. 2 (2013): 242–60.

———. "The Work of Wearing Cameras: Body-Worn Devices and Police Media Labor." In *The Routledge Companion to Labor and Media*, edited by R. Maxwell, 252–64. New York: Routledge, 2016.

Goodwin, Charles. "Professional Vision." *American Anthropologist* 96, no. 3 (1994): 606–33.

Gregory, Derek. "From a View to a Kill: Drones and Late Modern War." *Theory, Culture and Society* 28, no. 7–8 (2011): 188–215.

Guariglia, Matthew. *Police and the Empire City: Race and the Origins of Modern Policing in New York.* Chapel Hill, NC: Duke University Press, 2023.

Gürsel, Zeynep Devrim. *Image Brokers: Visualizing World News in the Digital Age of Circulation.* Berkeley: University of California Press, 2016.

Haggerty, Kevin D., and Ajay Sandhu. "The Police Crisis of Visibility." *IEEE Technology and Society Magazine* 33, no. 2 (2014): 9–12. https://doi.org/10.1109/MTS.2014.2319912.

Heartland Alliance. *Working Together toward Safer Communities: Reflections from READI Chicago*. Chicago: Heartland Alliance, 2022. https://www.heartlandalliance.org/wp-content/uploads/2021/09/READI-Chicago-Working-Together-Toward-Safer-Communities-small.pdf.

Heath, Brad. "Police Secretly Track Cellphones to Solve Routine Crimes." *USA Today*, August 23, 2015. https://www.usatoday.com/story/news/2015/08/23/baltimore-police-stingray-cell-surveillance/31994181/.

Henderson, Stephen F. "Fourth Amendment Time Machines (and What They Might Say about Police Body Cameras)." *Journal of Constitutional Law* 18, no. 3 (2016): 934.

Hill, Kashmir. *Your Face Belongs to Us: A Secretive Startup's Quest to End Privacy as We Know It*. New York: Penguin Random House, 2023.

Human Rights Watch. *Dark Side: Secret Origins of Evidence in US Criminal Cases*. New York: Human Rights Watch, 2018.

Hurston, Zora Neale. "The 'Pet Negro' System." *American Mercury* 56 (1943): 593–600.

Illinois v. Wardlow, No. 528 U.S. 119 (January 12, 2000).

Jerolmack, Colin, and Shamus Khan. "Talk Is Cheap: Ethnography and the Attitudinal Fallacy." *Sociological Methods and Research* 43, no. 2 (2014): 178–209.

Jones, James H. *Bad Blood: The Tuskegee Syphilis Experiment*. Rev. ed. New York: Free Press, 1993.

Kahn, Jeremy. "The Story of a Snitch." *The Atlantic*, April 2007. https://www.theatlantic.com/magazine/archive/2007/04/the-story-of-a-snitch/305703/.

Kaltheuner, Frederike, ed. *Fake AI*. Manchester, UK: Meatspace Press, 2021.

Katz, Jack. "Armor for Ethnographers." *Sociological Forum* 34, no. 1 (2019): 264–75.

Kerr, Orin S. "The Mosaic Theory of the Fourth Amendment." *Michigan Law Review* 111, no. 3 (2012): 311–54.

Khan, Shamus. "The Subpoena of Ethnographic Data." *Sociological Forum* 34, no. 1 (2019): 253–63.

Kneese, Tamara. *Death Glitch: How Techno-Solutionism Fails Us in This Life and Beyond*. New Haven, CT: Yale University Press, 2023.

Lageson, Sarah Esther. *Digital Punishment: Privacy, Stigma, and the Harms of Data-Driven Criminal Justice*. New York: Oxford University Press, 2020.

La Vigne, Nancy G., Samantha S. Lowry, Joshua A. Markman, and Allison M. Dwyer. *Evaluating the Use of Public Surveillance Cameras for Crime Control and Prevention*. Washington, DC: Urban Institute, September 2011.

Leaders of a Beautiful Struggle, Erricka Bridgeford, Kevin James v. Baltimore Police Department, Michael S. Harrison, No. 20-1495 (4th Cir. Ct. App. July 24, 2021).

Lee, Min Kyung, Daniel Kusbit, Evan Metsky, and Laura Dabbish. "Working with Machines: The Impact of Algorithmic and Data-Driven Management on Human Workers." In *Proceedings of the 33rd Annual ACM Conference on Human Factors in Computing Systems*, edited by B. Begole, J. Kim, K. Inkpen, and W. Wood, 1603-12. New York: ACM Press, 2015.

Leo, Richard A. "Trials and Tribulations: Courts, Ethnography, and the Need for an Evidentiary Privilege for Academic Researchers." *American Sociologist* 26 (1995): 113-34.

Leonardi, Paul M. "Materiality, Sociomateriality, and Socio-Technical Systems: What Do These Terms Mean? How Are They Different? Do We Need Them?" In *Materiality and Organizing: Social Interaction in a Technological World*, edited by Paul M. Leonardi, Bonnie A. Nardi, and Jannis Kallinikos, 24-48. New York: Oxford University Press, 2012.

Lilienfeld, Scott O., and Kristin Landfield. "Science and Pseudoscience in Law Enforcement: A User-Friendly Primer." *Criminal Justice and Behavior* 35, no. 10 (2008): 1215-30.

Lipsitz, George. *How Racism Takes Place*. Philadelphia: Temple University Press, 2011.

Litch, Michael. *Draft Technical Report for SECURES Demonstration in Hampton and Newport News, Virginia*. Washington, DC: U.S. Department of Justice, Office of Justice Programs, January 2011. https://www.ojp.gov/library/publications/draft-technical-report-secures-demonstration-hampton-and-newport-news-virginia.

Loewenstein, Antony. *The Palestine Laboratory: How Israel Exports the Technology of Occupation around the World*. New York: Verso, 2023.

Lowenkamp, Christopher T., Marie VanNostrand, and Alexander Holsinger. *The Hidden Costs of Pretrial Detention*. Houston: Laura and John Arnold Foundation, 2013.

Lubet, Steven. *Interrogating Ethnography: Why Evidence Matters*. New York: Oxford University Press, 2018.

Lucy Parsons Labs. "ShotSpotter Creates Thousands of Unfounded Police Deployments, Fuels Unconstitutional Stop-and-Frisk, and Can Lead to False Arrests." MacArthur Justice Center, July 21, 2022. https://endpolicesurveillance.com/.

Lyon, David. "9/11, Synopticon, and Scopophilia: Watching and Being Watched." In *The New Politics of Surveillance and Visibility*, edited by Kevin D. Haggerty and Richard V. Ericson, 35–54. Toronto: University of Toronto Press, 2006.

Malka, Adam. *The Men of Mobtown: Policing Baltimore in the Age of Slavery and Emancipation*. Chapel Hill: University of North Carolina Press, 2018.

Mann, Steve, Jason Nolan, and Barry Wellman. "Sousveillance: Inventing and Using Wearable Computing Devices for Data Collection in Surveillance Environments." *Surveillance and Society* 1, no. 3 (2003): 331–55.

Mathiesen, Thomas. "The Viewer Society: Michel Foucault's 'Panopticon' Revisited." *Theoretical Criminology* 1, no. 2 (1997): 215–34.

McVeigh, Karen. "US Drug Agency Surveillance Unit to Be Investigated by Department of Justice." *The Guardian*, August 6, 2013, sec. World. https://www.theguardian.com/world/2013/aug/06/justice-department-surveillance-dea.

Mehrotra, Dhruv, and Joey Scott. "Here Are the Secret Locations of ShotSpotter Gunfire Sensors." *Wired*, February 22, 2024. https://www.wired.com/story/shotspotter-secret-sensor-locations-leak/.

Merton, Robert K. "The Unanticipated Consequences of Purposive Social Action." *American Sociological Review* 1, no. 6 (1936): 894–904.

Michel, Arthur Holland. *Eyes in the Sky: The Secret Rise of Gorgon Stare and How It Will Watch Us All*. New York: Houghton Mifflin Harcourt, 2019.

Molotch, Harvey. *Against Security: How We Go Wrong at Airports, Subways, and Other Sites of Ambiguous Danger*. Princeton, NJ: Princeton University Press, 2012.

Morozov, Evgeny. *To Save Everything, Click Here: The Folly of Technological Solutionism*. New York: Public Affairs, 2013.

Morral, Andrew R., Terry L. Schell, Brandon Crosby, Rosanna Smart, Rose Kerber, and Justin Lee. *Preliminary Findings from the Aerial Investigation Research Pilot Program*. Santa Monica, CA: RAND, January 27, 2021.

Morral, Andrew R., Terry L. Schell, Rosanna Smart, Brandon Crosby, Justin W. Lee, and Rose Kerber. *Evaluating Baltimore's Aerial Investigation Research Pilot Program: Interim Report*. Santa Monica, CA: RAND, 2021. https://www.rand.org/pubs/research_reports/RRA1131-2.html.

Moy, Laura. *Complaint for Relief against Unauthorized Radio Operation and Willful Interference with Cellular Communications: Petition for an Enforcement*

Advisory on Use of Cell Site Simulators by State and Local Government Agencies. Washington, DC: Georgetown University Law Center, August 16, 2016. https://www.documentcloud.org/documents/3013988-CS-Simulators-Complaint-FINAL.html#document/p4/a314050.

Muhammad, Khalil Gibran. *The Condemnation of Blackness: Race, Crime, and the Making of Modern Urban America.* 2nd ed. Cambridge, MA: Harvard University Press, 2019.

Muñiz, Ana. *Borderland Circuitry: Immigration Surveillance in the United States and Beyond.* Oakland: University of California Press, 2022.

Nader, Laura. "Up the Anthropologist: Perspectives Gained from Studying Up." ERIC Clearinghouse, 1972. https://eric.ed.gov/?id=ED065375.

Nagin, Daniel S. "Deterrence in the Twenty-First Century." *Crime and Justice* 42, no. 1 (2013): 199–263.

Narayanan, Arvind, and Sayash Kapoor. *AI Snake Oil: What Artificial Intelligence Can Do, What It Can't, and How to Tell the Difference.* Princeton, NJ: Princeton University Press, 2024.

National Policing Institute. *A Review of the Baltimore Police Department's Use of Persistent Surveillance (Baltimore Community Support Program).* Arlington, VA: National Policing Institute, January 30, 2017.

Newell, Bryce Clayton. "Context, Visibility, and Control: Police Work and the Contested Objectivity of Bystander Video." *New Media and Society* 21, no. 1 (2019): 60–76.

———. *Police Visibility: Privacy, Surveillance, and the False Promise of Body-Worn Cameras.* Berkeley: University of California Press, 2021.

Newman, Lily Hay. "How Baltimore Became America's Laboratory for Spy Tech." *Wired*, September 4, 2016. https://www.wired.com/2016/09/baltimore-became-americas-testbed-surveillance-tech/.

Noble, Safia Umoja. *Algorithms of Oppression: How Search Engines Reinforce Racism.* New York: NYU Press, 2018.

Ohm, Paul. "The Many Revolutions of Carpenter." *Harvard Journal of Law and Technology* 32, no. 2 (2019): 358–416.

O'Neil, Catherine. *Weapons of Math Destruction: How Big Data Increases Inequality and Threatens Democracy.* New York: Crown Random House, 2016.

Palys, Ted, and John Lowman. "Anticipating Law: Research Methods, Ethics, and the Law of Privilege." *Sociological Methodology* 32 (2002): 1–17.

———. *The Belfast Project Autopsy: Who Can You Trust?* London: Routledge, 2016.

———. "Defending Research Confidentiality 'to the Extent the Law Allows:' Lessons from the Boston College Subpoenas." *Journal of Academic Ethics* 10 (2012): 271–97.

———. *Informed Consent, Confidentiality and the Law: Implications of the Tri-Council Policy Statement*. Burnaby, BC: Simon Fraser University, January 31, 1999. http://www.sfu.ca/~palys/Conf&Law.html.

———. *Protecting Research Confidentiality: What Happens When Law and Ethics Collide*. Toronto: James Lorimer, 2014.

Pfeffer, Jeffrey, and Gerald R. Salancik. *The External Control of Organizations: A Resource Dependence Perspective*. Stanford, CA: Stanford University Press, 2003.

Power, Garrett. "Apartheid Baltimore Style: The Residential Segregation Ordinances of 1910–1913." *Maryland Law Review* 42, no. 2 (1983): 289–329.

Prince Jones v. United States, No. 15-CF-322 (DC Ct. App. September 21, 2017).

Raji, Inioluwa Deborah, I. Elizabeth Kumar, Aaron Horowitz, and Andrew Selbst. "The Fallacy of AI Functionality." In *FAccT '22: Proceedings of the 2022 ACM Conference on Fairness, Accountability, and Transparency*, June 2022, 959–72. https://doi.org/10.1145/3531146.3533158.

Reed, Isaac Ariail. *Interpretation and Social Knowledge: On the Use of Theory in the Human Sciences*. Chicago: University of Chicago Press, 2011.

Reel, Monte. "Secret Cameras Record Baltimore's Every Move from Above." *Bloomberg Businessweek*, August 23, 2016. https://www.bloomberg.com/features/2016-baltimore-secret-surveillance/.

Richards, Neil. *Why Privacy Matters*. New York: Oxford University Press, 2021.

Rios, Victor M. *Punished: Policing the Lives of Black and Latino Boys*. New York: NYU Press, 2011.

Rizer, Arthur. "Very Little Stands between the U.S. and a Technological Panopticon." *Slate*, November 19, 2020. https://slate.com/technology/2020/11/law-enforcement-facial-recognition-technology.html.

Rocah, David R., Ashley Gorski, Brett Max Kaufman, Alexia Ramirez, Nathan Freed Wessler, and Ben Wizner, Leaders of a Beautiful Struggle, Erricka Bridgeford, Kevin James v. Baltimore Police Department, Michael S. Harrison—Complaint for Declaratory and Injunctive Relief, No. 20-929 (Dis. Ct. of Md. April 9, 2020).

Roediger, David R. *Working toward Whiteness: How America's Immigrants Became White*. New York: Basic Books, 2006.

Rothstein, Richard. *The Color of Law: A Forgotten History of How Our Government Segregated America*. New York: W. W. Norton, 2017.

Ruiz, Andres. "The Belfast Project: How an American University Almost Started a Civil War." *StMU Research Scholars* (blog), November 30, 2021. https://stmuscholars.org/the-belfast-project-how-an-american-university-almost-started-a-civil-war/.

Sankin, Aaron, Dhruv Mehrotra, Surya Mattu, and Annie Gilbertson. "Crime Prediction Software Promised to Be Free of Biases; New Data Shows It Perpetuates Them." *The Markup*, December 2, 2021.

Scannell, R. Joshua. "This Is Not Minority Report: Predictive Policing and Population Racism." In *Captivating Technology: Race, Carceral Technoscience, and Liberatory Imagination in Everyday Life*, edited by Ruha Benjamin, 107–29. Durham, NC: Duke University Press, 2019.

Scarce, Rik. "(No) Trial (but) Tribulations: When Courts and Ethnography Conflict." *Journal of Contemporary Ethnography* 23, no. 2 (1994): 123–49.

Schuppli, Susan. *Material Witness: Media, Forensics, Evidence*. Cambridge, MA: MIT Press, 2020.

Schwartz, Louis-Georges. *Mechanical Witness: A History of Motion Picture Evidence in U.S. Courts*. New York: Oxford University Press, 2009.

Seaver, Nick. "Studying Up: The Ethnography of Technologists." *Ethnography Matters* (blog), March 10, 2014. https://ethnographymatters.net/blog/2014/03/10/studying-up/.

Seim, Josh. "Participant Observation, Observant Participation, and Hybrid Ethnography." *Sociological Methods and Research*, 2021. https://doi.org/10.1177/0049124120986209.

Sharkey, Patrick. *Uneasy Peace: The Great Crime Decline, the Renewal of City Life, and the Next War on Violence*. New York: W. W. Norton, 2018.

Sharkey, Patrick, and Alisabeth Marsteller. "Neighborhood Inequality and Violence in Chicago, 1965–2020." *University of Chicago Law Review* 89, no. 2 (2022): 3.

Shestakofsky, Benjamin. *Behind the Startup: How Venture Capital Shapes Work, Innovation, and Inequality*. Oakland: University of California Press, 2024.

Simpson, J. Cavanaugh. "Prying Eyes: Military-Grade Surveillance Keeps Watch Over Baltimore and City Protests, but Catches Few Criminals." *Baltimore Magazine*, August 5, 2020. https://www.baltimoremagazine.com/section/community/surveillance-planes-watch-over-baltimore-but-catch-few-criminals/.

Slobogin, Christopher. *Virtual Searches: Regulating the Covert World of Technological Policing*. Oakland: University of California Press, 2022.

Snyder, Benjamin H. "'All We See Is Dots': Aerial Objectivity and Mass Surveillance in Baltimore." *History of Photography* 45, no. 3–4 (2021): 376–87. https://doi.org/10.1080/03087298.2022.2108263.

———. "'Big Brother's Bigger Brother': The Visual Politics of (Counter) Surveillance in Baltimore." *Sociological Forum* 35, no. 4 (2020): 1315–36.

Soderberg, Brandon. "Baltimore Defense Attorneys Claim Surveillance Plane Footage Contradicts Law Enforcement Account of Police Shooting." *The Appeal*, February 13, 2020. https://theappeal.org/baltimore-police-shooting-new-motion/.

———. "Persistent Transparency: Baltimore Surveillance Plane Documents Reveal Ignored Pleas to Go Public." *Baltimore City Paper*, November 1, 2016. https://www.citypaper.com/news/mobtownbeat/bcp-110216-mobs-aerial-surveillance-20161101-story.html.

Soss, Joe, and Vesla Weaver. "Police Are Our Government: Politics, Political Science, and the Policing of Race-Class Subjugated Communities." *Annual Review of Political Science* 20 (2017): 565–91.

Spivack, Jameson. *Cop Out: Automation in the Criminal Legal System*. Washington, DC: Georgetown Law Center on Privacy and Technology, March 23, 2023. https://copout.tech/wp-content/uploads/2023/03/CPT_Cop-Out_Essay.pdf.

Stone, Geoffrey R. "Discussion—Above the Law: Research Methods, Ethics, and the Law of Privilege." *Sociological Methodology* 32 (2002): 19–27.

Stuart, Forrest. "Constructing Police Abuse after Rodney King: How Skid Row Residents and the Los Angeles Police Department Contest Video Evidence." *Law and Social Inquiry* 36, no. 2 (2011): 327–53.

———. *Down, Out, and Under Arrest: Policing and Everyday Life in Skid Row*. Chicago: University of Chicago Press, 2016.

Taylor, Emmeline. "Lights, Camera, Redaction . . . Police Body-Worn Cameras: Autonomy, Discretion, and Accountability." *Surveillance and Society* 14, no. 1 (2016): 128–32.

Theodos, Brett, Eric Hangen, and Brady Meixell. "The Black Butterfly: Racial Segregation and Investment Patterns in Baltimore." Urban Institute, February 5, 2019. https://apps.urban.org/features/baltimore-investment-flows/.

Turner, Fred. *From Counterculture to Cyberculture: Stewart Brand, the Whole Earth Network, and the Rise of Digital Utopianism*. Chicago: University of Chicago Press, 2006.

U.S. Department of Justice. *Investigation of the Baltimore City Police Department*. Washington, DC: U.S. Department of Justice, Civil Rights Division, August 10, 2016.

Van Maanen, John. "The Moral Fix: On the Ethics of Fieldwork." In *Contemporary Field Research: A Collection of Readings*, edited by R. M. Emerson, 269–87. Boston: Little Brown, 1983.

Vertesi, Janet. *Seeing Like a Rover: How Robots, Teams, and Images Craft Knowledge of Mars*. Chicago: University of Chicago Press, 2014.

Villa-Nicholas, Melissa. *Data Borders: How Silicon Valley Is Building an Industry around Immigrants*. Oakland: University of California Press, 2023.

Vinsel, Lee. "You're Doing It Wrong: Notes on Criticism and Technology Hype." *STS News* (blog), February 1, 2021. https://sts-news.medium.com/youre-doing-it-wrong-notes-on-criticism-and-technology-hype-18b08b4307e5.

Virilio, Paul. *War and Cinema: The Logistics of Perception*. London: Verso, 2009.

Vogel, Ed, and Fletcher Nickerson. "Surveillance Technologies Don't Create Safety; They Intensify State Violence." *Truthout*, May 21, 2023. https://truthout.org/articles/surveillance-technologies-dont-create-safety-they-intensify-state-violence/.

Waldman, Ari Ezra. *Industry Unbound: The Inside Story of Privacy, Data, and Corporate Power*. Cambridge: Cambridge University Press, 2021.

Washington, Harriet A. *Medical Apartheid: The Dark History of Medical Experimentation on Black Americans from Colonial Times to the Present*. New York: Vintage, 2008.

WBEZ Chicago. *Motive* (podcast), season 5, January–March 2023. https://www.wbez.org/shows/motive/.

Weinstein, Jack B., and Catherine Wimberly. "Secrecy in Law and Science." *Cardozo Law Review* 23, no. 1 (2001): 1–32.

Weisburd, David, Rosann Greenspan, Edwin E. Hamilton, Kellie A. Bryant, and Hubert Williams. *Abuse of Police Authority: A National Study of Police Officers*. Washington, DC: Police Foundation, 2001.

Williams, Jennie K. "Trouble the Water: The Baltimore to New Orleans Coastwise Slave Trade, 1820–1860." *Slavery and Abolition* 41, no. 2 (2020): 275–303.

Woods, Baynard, and Brandon Soderberg. *I Got a Monster: The Rise and Fall of America's Most Corrupt Police Squad*. New York: St. Martin's Press, 2020.

Zomorodi, Manoush, and Alex Goldmark. "Eye in the Sky." *Radiolab* (podcast), WNYC Studios, June 18, 2015. https://www.wnycstudios.org/story/eye-sky.

Zuboff, Shoshana. *The Age of Surveillance Capitalism: The Fight for a Human Future at the New Frontier of Power*. New York: Public Affairs, 2019.

Index

ACLU (American Civil Liberties Union), 12, 130, 168, 218, 232
AcuraLink, 159-160
Adams, Eric, 10-11
AI (artificial intelligence), 206
AIR (Aerial Investigation Research), vii, 25. *See also* spy plane program
algorithms, ix, 54, 209, 211
ALPR (Automated License Plate Reader). *See* LPR
Altman, Sam, 206
American Sociological Association, 239
Anderson, Chris, 206
Angel Fire, 23
Arnold Ventures, 23, 228
Arnold, John, 23-24. *See also* Arnold Ventures
Arsiniega, Brittany, 134
Avedon, Richard, 190
Axon, 17

Baltimore: Camden Yards, 96; Carrollton Ridge, 89; Eastside, 105; Federal Hill, 89; Greektown, 99-100; homicide trends in, 7, 89-90; Lexington Market, 96; McElderry Park, 88; Mondawmin Mall, 108, 110; Penn North, 125; Sandtown, 69-71, 86, 125; Westside, 25, 69, 105, 108. *See also* Black Butterfly
Baltimore Sun, coverage of surveillance technology in, 25, 98, 103, 109-110
Bates, Ivan, 123, 135-138
Belfast Project, The, 235
Bell, Monica, 92
Benjamin, Ruha, 16
Bentham, Jeremy, 12
Big Brother, 8, 56, 95, 109, 124, 147, 173
Black Butterfly, 87-89, 144
Black Panthers, 124
Bloomberg Businessweek, 25
blue light cameras, 100-101
body cams, 17, 127, 134
boomerism, 9-11, 45, 99, 152, 205, 210, 224-225
borderlands, 71, 75, 87-88, 96-98, 105, 109, 113, 120
Boston College, 235

BPD (Baltimore Police Department): history of, 92–93, 251n18; organizational dysfunction of, 141; public distrust in, 33, 36, 85, 126–128; racism of, 129–130
Brady violations, 197–199
Brayne, Sarah, 54, 208–209, 215, 264n14
Brown, Lawrence, 87

Carpenter v. United States, 164, 168–172, 260n19
Carpenter, Timothy, 169–172, 182. See also *Carpenter v. United States*
Carroll County State's Attorney, 137–138
CBT (cognitive-behavioral therapy), 220
CCOPS (Community Control Over Police Surveillance), 218
CCTV (Closed Circuit Television), 38, 40, 71, 76. See also CitiWatch
cell site simulator. See stingray (technology)
Center for Investigative Reporting, 113
Center on Privacy and Technology, 211
Chicago, 210
chilling effect, 147
CitiWatch: 24, 38–40, 71, 101–106, 253n53; effectiveness of, 40, 107–111. See also Video Patrol program
Citizen app, 204
Ciudad Juárez, Mexico, 5, 52
ClearView AI, 210
Cohen, Julie, 174
Community With Solutions, 126
cop-watching. See sousveillance

COVID-19, vii–x, 19, 61, 74, 184, 228, 232
creepiness, 12, 67, 103, 110, 151, 155, 173–174
Crime Stoppers, 224, 267n47
criti-hype, 13, 110, 173–176, 205
Cruise, Tom, 207
CSLI (Cell Site Location Information), 114–115, 169
Cusumano v. Microsoft Corporation, 236
Cusumano, Michael, 236

DARPA (Defense Advanced Research Projects Agency), 21
Dayton, Ohio, 19
defense attorneys, 83, 142, 181–183, 199
deterrence effect, 35, 93, 103, 105
Detroit, 211
DHS (Department of Homeland Security), 101
discriminatory design, 16
District Court of Maryland, 168
Dixon, Sheila, 109
DOD (Department of Defense), 132
DOJ (Department of Justice), 127
doomerism, 12, 45, 152, 173, 205–206, 210, 224–225
Downtown Partnership, 94–96
dragnet surveillance, 130, 208–209
drones, 3, 22, 157
Du Bois, W.E.B., 227
dystopia, 9–13, 129, 173–176, 210, 213

EFF (Electronic Frontier Foundation), 12
Enemy of the State, 22, 51
ethnography: and academic freedom, 236–237, and Black trauma, 229–231;

and certificates of confidentiality, 237; and confidentiality, 237–239; and consent, 230–231; ethics of, 229–231; First Amendment rights in, 236–237; and limited confidentiality policies, 234, 238–239; positionality in, 227, 229; and shaping the data, 231–234; and "studying up," 227, 232; and subpoena of field notes, 234–238, 269n16

experimentation: ethical forms of, 218–223; harms of, 15–16, 82–86, 116–117, 162, 168, 173–176, 185–186, 199–202, 213; problem of, ix, 9, 15–16, 61, 65, 84–86, 107–109, 111, 117–121, 147, 186, 202, 207–214, 224–225; regulation of, 214–218

Eyes in the Sky (Michel), 22

Facebook, 61, 86, 204
facial recognition technology, viii, 12, 48–49, 210–211
"fake it 'til you make it," 21, 83, 215
Fallujah, Iraq, 23
false positives, 57, 74–80, 113, 192–194, 211, 222
fast failure, 21, 215
FBI (Federal Bureau of Investigations), 115, 117
Floyd, George, 124
focused deterrence, 219, 267n39
FOIA (Freedom of Information Act), 17, 83, 209
forensics, 145, 182
forming cause, 89, 120
Foucault, Michel, 12, 151
Fourth Amendment, 94, 164, 168–172, 182, 198, 204

Fourth Circuit Court of Appeals, 168–172, 182, 198, 233, 259n14
"fruit of the poisonous tree," 183, 234

Gates, Bill, 206
Gertner, Nancy, 196
Goodwin, Charles, 188
Google Earth, 34, 41, 44
Gorsuch, Neal, 260n19
GPS (Global Positioning System), 46, 123, 159–161
Gray, Freddie, 108
Greektown Community Development Corporation, 99
Gregory, Roger, 169, 172
ground cameras, 34, 48, 72, 74. *See also* CCTV
GTTF (Gun Trace Task Force), 122–123, 135–138
Guariglia, Matthew, 134
gun violence, 89, 120, 218–219
gunshot detection systems, ix, 17, 111–112. *See also* ShotSpotter

Harris Technologies, 115, 117, 157
Harrison, Michael, 61–64, 82, 153–154
Holliday, George, 188
Human Rights Watch, 197
hype, 10, 13, 141, 152, 205. *See also* criti-hype

impact assessment, 216
inequality, 14, 17, 85–86, 117–118, 175–176
infallible state assumption, 10
Instagram, 36, 204
"internet of things," 159
Iraq, 21

IRB (Institutional Review Board), 235, 239
iView, 28, 39

Johns Hopkins University, 112–113
Johnson, Brandon, 210
Jones, Joyous, 125–134, 143–144, 148

Kerr, Orin S., 260n20
King, Rodney, 188–189

LAPD (Los Angeles Police Department), 207
Laura and John Arnold Foundation. *See* Arnold Ventures
lateral surveillance, 221
law of crime concentration, 220
Leaders of a Beautiful Struggle v. Baltimore Police Department, 168–172, 182, 198, 232–233, 259n15
Lipsitz, George, 91
Lowman, John, 235
LPR (License Plate Reader), 69, 73, 165–167, 175

MacArthur Justice Center, 209
McNutt, Ross, 21–22
Meta. *See* Facebook
metadata, 114, 116
Michel, Arthur Holland, 22
Microsoft, 236
Minority Report, 207
mission creep, 103, 115, 121, 157
MIT (Massachusetts Institute of Technology), 21
Mosby, Marilyn, 177–178
"move fast and break things," 9, 21, 155

Moy, Laura, 117

network analysis, 161–162
1984 (Orwell), 9
NYPD (New York Police Department), 211

O'Neil, Cathy, 83
objectivity, 62, 68, 135, 139, 187–189, 192–196, 200–202
observant-participation, 244n34
Ohm, Paul, 171
OPD (Office of the Public Defender), 181–182
Orwell, George, 9

Palantir, 208, 264n14
Palys, Ted, 235
pandemic. *See* COVID-19
panopticism, 151
panopticon, 12, 67, 99, 111
parallel construction, 115, 196–200
participant-observation, 20. *See also* observant-participation
pattern-of-life analysis, 7
peer monitoring. *See* lateral surveillance
Photoshop, 211
Planning Systems Incorporated, 112
Police Foundation, 25
policing: abolition of, 130; and "bad apple" officers, 63, 83, 201, 212; broken windows, 94; discretionary power of, 54; evasion of, 45; evidence-based, 61, 215–216, 221; hot spots, 105, 207; misconduct in, 122–123, 127, 134, 136–138; and overtime abuse, 142; and supercitizens, 134; and supercops, 9, 15, 176, 212; and superprivacy, 134–135,

137–139, 227; university police, 112; and unwarranted contact, 77–78, 107
Policing Project, 19, 82, 228, 232–233
Popular Science, 12
PowerPoint, 61
predictive policing. *See* PredPol
PredPol, 17, 207–208
pretrial detention, 184, 199–200, 211
privacy: erosion of, 101–102; invasion of, 12, 35, 47, 62, 77–78, 94–95, 133–134, 154, 165–166, 168–175; performance of, 64, 121, 164–167, 171, 174–175
privacy harms. *See* experimentation, harms of
privacy law. *See* Fourth Amendment
privatization, 111, 114, 119–121
Project Green Light, 211
property crime, 107–108
prosecutors, 83, 116, 142, 183
pseudoscience, 15, 61, 84, 107, 119, 247n3
PSS (Persistent Surveillance Systems), 21–25
public health approaches to gun violence reduction, 133, 219–223
public housing, 69, 162
public-private partnerships: ix, 111, 114, 201; and profit motive, 83, 119–120, 167, 201, 213; regulation of, 214–218. *See also* public-public partnerships
public-public partnerships, 219–220
Pulaski effect. *See* false positives

race-class subjugated communities, 86, 88–89, 175, 211, 222
Radiolab, 18, 23
RAND Corporation, 19, 53
randomized controlled trials, 223

rapid prototyping, 21, 74, 159, 175, 215. *See also* RPD
Rawlings-Blake, Stephanie, 109
READI (Rapid Employment And Development Initiative), 219–222
reciprocal silence, 134. *See also* snitching
Reel, Monte, 25
Richards, Jawan, 122–124, 135–138
risk, 67, 84–85, 116, 119, 145–146
Roberts, John, 170–172
RPD (Rapid Product Development), 21

safety, feeling of, 11, 92, 95–96
SAO (State's Attorney's Office), 180
Schlanger, Elliot, 107
Schmoke, Kurt, 95
SECURES system, 112
security cameras. *See* ground cameras
segregation, 87–92, 119–120. *See also* Black Butterfly
ShotSpotter, 17, 113–114, 209–210, 254n72, 265n15. *See also* gunshot detection systems
Silicon Valley, 10–11, 21, 111
Simmons Memorial Baptist Church, 125, 143, 204
Simmons, Duane, 125
Slate, 12
Smith, Will, 22
snitching, 36, 134, 146
socio-technical systems, 55. *See also* surveillance technology, organizational shaping of
soda straw problem, 22
Soderberg, Brandon, 135, 141
solutionism, ix, 10, 20–21, 129, 226
SoundThinking. *See* ShotSpotter
sousveillance, 124, 128–129, 141–143, 203

spy plane program: effectiveness of, 35–37, 50–52, 81–82, 267n48; and Four-Day Rule, 163–168, 171; ineffectiveness of, 53–55, 163, 194, 247n11; privacy policy of, 153–155, 167, 198, 259n7; public opinion of, 126, 133–134, 145, 248n5, 257n19, 259n15; racial bias in, 63–64, 77, 85–86, 202; and supplemental surveillance, 158–163, 165–167, 197–198, 233; technical details of, 2–4, 28–29; tracking work in, 2–7, 30–34, 37–52, 69–82, 145–147, 159–164; and 2016 Baltimore trial, 23–25, 135; unconstitutionality of, 47, 154, 155–156, 159–163, 164–172, 182–185, 204, 259n15; use in court, 137–143, 181–187, 190–200

St. Louis, 148, 216

stingray (technology), 114–118, 162, 181, 186

stop-and-frisk, 131

structural racism, 63

surveillance technology: bias in, 16, 109, 208–209; community control over, 124, 216–219; and corruption, 135–138, 192–196, 210–213; deployment of, 16, 21, 59, 117, 215; glitchy and ineffective, 40, 46, 74–75, 82, 100–101, 113, 119, 175, 208, 210–213; and linking multiple technologies, 166; magic of, 52–53, 55, 99, 171–174; and mistakes, 43, 57, 67–69, 74–83, 116, 212; organizational shaping of, 141, 147, 158, 167, 179–180, 201–202, 213

Surveillance Technology Oversight Program, 211

technology start-ups, 83, 155–156, 164, 204, 206–207, 215

terrorism, 102–103, 115

The Wire, 229

time machine problem, 169–172

time travel, 4, 27, 42–43, 162, 165–166, 169–172

Toledo, Adam, 210

transparency, ix, 17, 19, 24, 60–63, 113, 115, 119, 216–218

Troubles, The, 235

true crime entertainment, 230

trust, 35, 53–54, 85, 126, 220

Twitter. *See* X (social media platform)

U.S. Border Patrol, 68

University of Baltimore, 19

Urban Institute, 108–109

USA Today, 115–116

utopia, 11, 129, 132, 174, 222

video evidence, 140, 187–190

Video Patrol program, 94–99, 205

vigilantism, 204, 225

Vinsel, Lee, 13, 205

violence as a disease, 223. *See also* public health approaches to gun violence reduction

violence interrupters, 133, 147–148

violent crime, 107–108, 119

violent repeat offenders, 129

virtual search, 169–170

Waldman, Ari, 164

WAMI (Wide Area Motion Imagery): and blurry, pixelated imagery, 5–6, 28, 40, 62, 65, 140–141, 156, 192; brittleness of, 190–196, 200–202, 210, 212; interpretation of 59, 63, 68, 72, 76, 139–141, 180, 200–202; limitations of, 32, 53–54, 65, 73–76, 140, 156–157, 173–174, 190–196;

origins of, 22–23, 245n38; and resolution of imagery, 6, 28, 153, 156, 223, 153; technical aspects of, 3–4, 28–29, 65

warrants, 117

Watch Center, 102

White L, 87–89, 144, 227. *See also* Black Butterfly

White Spatial Imaginary, 91, 96, 99, 105, 109, 119–120

Wigmore test, 237, 269n19

Wilkinson, J. Harvie, 259n15

Williams, Archie, 125–134

Williams, Robert, 211

Wired, 12, 111

witnessing, 35, 52, 54, 187–190

Wong, Matteo, 206

X (social media platform), 204

Yoffie, David, 236

YouTube, 53, 204

Founded in 1893,
UNIVERSITY OF CALIFORNIA PRESS
publishes bold, progressive books and journals
on topics in the arts, humanities, social sciences,
and natural sciences—with a focus on social
justice issues—that inspire thought and action
among readers worldwide.

The UC PRESS FOUNDATION
raises funds to uphold the press's vital role
as an independent, nonprofit publisher, and
receives philanthropic support from a wide
range of individuals and institutions—and from
committed readers like you. To learn more, visit
ucpress.edu/supportus.